DR KARL'S
SURFING SAFARI
THROUGH SCIENCE

Dr Karl Kruszelnicki

Illustrated by Pilar Costabal
Designed by Lisa Reidy

ABC
BOOKS

CONTENTS

SURF'S UP!

IN 2020,
THE TSUNAMI
OF COVID-19
SWEPT THE WORLD ...

The year 2020 has reminded us that science is a bit like a wave. By that, I mean that science is a dynamic process that ebbs and flows, rather than something set in stone.

The COVID-19 pandemic has scientists working hard to find the best way to save the greatest number of people from terrible health outcomes – in an ever-shifting environment.

As I write, the world's most developed nation, the USA, has the highest number of COVID-19 cases and deaths. This is partly due to the flood of conspiracy theories and disinformation awash on the internet and in the media.

Amidst the panic and false information, Dr Anthony Fauci, the Director of the National Institute of Allergy and Infectious Diseases in the US, has been a calm voice of reason, tirelessly advocating that we must rely on current expert scientific and medical knowledge in our efforts to manage the virus. Doctor Fauci describes science as being an 'attempt, in good faith, to get to the facts'. What he means is that science is NOT a collection of facts – that's what you have an encyclopaedia for. Instead, science is a process of *discovering* facts through curious exploration, and then using them to understand the Universe around us – which is precisely what we're trying to do with COVID-19.

Science is an ongoing process. It's self-correcting – which, let me emphasise, is a strength, not a weakness. So, as the pandemic plays out, we may find that the most up-to-date advice will be different from the initial advice. That might sound confusing at first, but scientists are flexible and willing to look at results and tweak the advice accordingly.

The real game-changer in this pandemic will be developing a vaccine or drugs to effectively prevent or treat the disease. In the meantime, all thanks must go to the research scientists and medical staff who, despite personal health risks, are doing everything they can to get us to a better place.

Now, let's go catch a wave. Did you know, it's a process, not a thing?

Yo, dudes. I've paddled onto the wave of innovation to splash loads of pop-up holograms all over this book for you, and I'm stoked!

Yep, just like my previous book, *Dr Karl's Random Road Trip Through Science*, every chapter in this book has gnarly pop-up holograms (or 'augmented reality') as a bonus.

All you need is a device and Net access. But only do this on dry land – your phone might be waterproof, but this book ain't (wipeout, bro)!

HERE'S HOW IT WORKS:
1. Download the free and modestly named Dr Karl app
2. Allow camera access, then click 'BEGIN'.
3. Download updates.
4. Hold your device/smartphone about 30 cm above a trigger page (the book cover or first page of any chapter) – make sure your volume is turned up.
5. You should see a flattened image of me talking to you. Adjust the angle of the phone until the 'Mini-Me' hologram looks 3D.
6. After I pop up out of the page to give you a rad intro, you'll notice 'buttons' under my feet. Press them to surf an ocean of awesome information – videos, words, gifs, etc.

Like, totally check out drkarl.com for a video demo, man – it'll be epic!

5

COFFEE –
GRINDING THE PERFECT CUP

Coffee contains caffeine, the world's most popular legal drug. We have been sipping it as a beverage since at least the middle of the 15th century. Legend has it that the stimulant effects of coffee were first discovered in Yemen or Ethiopia by a goatherd, who got curious when he saw his goats getting frisky after eating some beans off a coffee plant.

You'd think that, after six centuries, we would well and truly have locked in the elusive secret of making a consistent and high-quality cup of coffee. But, surprisingly, no. According to an international consortium studying the matter, we have been doing it the wrong way all this time.

HIT ME WITH YOUR BEST SHOT

Let me lay it on the line. For me, instant coffee is not coffee. And neither is plunger nor drip filter coffee (which is a bit better than instant, but is still not 'real' coffee). I'm talking about the purest form of coffee – totally unadulterated by lactose-free milk, fructose-free sugar or (heaven help us) citrus. I am talking espresso.

Italy is the home of the espresso. The Italian definition of 'espresso', according to the Instituto Nazionale Espresso Italiano, states that you force hot water through 6.5–7.5 grams of ground coffee, at a temperature of 86–90°C, and at a pressure of some 9 atmospheres (atm). Do this for 20–30 seconds and you should end up with 23–27 mL of hot, dark, coffee liquid.

The American Specialty Coffee Association has a similar definition of espresso, with a similar water pressure of 9–10 atm and the same flow time of 20–30 seconds, but using slightly more ground coffee (7–9 grams) and a slightly higher water temperature (92–95°C). This gives you 25–35 mL of beverage.

But your average coffee shop is not a stickler for these guidelines. They will start with a greater mass of ground coffee (say, 15–22 grams), which gives you a larger volume of steaming-hot dark liquid (say, 25–50 mL).

Coffee & $$$

In 2015, in the USA, the coffee industry provided more than 1.5 million jobs. That accounted for 1.6% of the US Gross Domestic Product (US$225.2 billion) and about US$30 billion in taxes.

THE DEVIL IS IN THE CONSISTENCY

Baristas don't necessarily follow these rules, but they are very particular about their methods. They try really hard to deliver constantly high-quality espresso. But they can't do it! There is always a problem with consistency.

In some professional barista competitions, contestants are asked to make four espressos – but to their chagrin, each one will taste different. There is something 'uncontrolled' popping up in the espresso system.

That makes sense, because this system is actually very complex. There are so many things to consider:

- How old should the coffee beans be when you grind them?
- How finely should you then grind those beans?
- How much ground coffee should you use, and how firmly should you tamp it down?
- What temperature and pressure should the water that travels through the ground coffee be at – and for how long should you let the water come through?

That is a lot of variables. So let's make it manageable, and start with a single grain of finely ground coffee, nestled in between a whole bunch of other coffee grains, all packed into the coffee basket. In total, grains like this one take up about 82% of the available space, which means that about 18% is free for the pressurised hot water to force its way through. The grains vary enormously in size – 10–1,000 μm, or micrometres, in diameter. (For comparison, a shaft of hair on your scalp is about 50–70 μm.) That's a huge range – about 1 to 100. However, about 99% of the grains are less than 100 μm across. These smaller grains account for about 80% of the total surface area of all the grains of coffee. Each individual grain of ground coffee contains (at last count) some 2,000+ recognised chemicals.

Those chemicals have to leach out from inside each grain into the pressurised hot water being forced past it. This hot water starts as pure hot water at the top of the basket – but on its way down through the basket, the water picks up many different chemicals from the grains that it pushes past. This chemical-laden solution ends up at the bottom of the espresso basket – and finally, it is poured into your small cup to give you some heavenly goodness.

GREEN ROOM

SCIENCE IMPROVES COFFEE

A coffee consortium was put together to solve the Coffee Consistency Problem and explain why a barista using the exact same technique twice should end up with two espressos that taste quite different. This eclectic bunch of people with widely varying skills included a Melbourne barista (yep, Melbourne baristas are top-notch), computational chemists (computational chemistry marries chemistry with computer modelling) and mathematicians from Australia, Switzerland, the United Kingdom, Ireland and the USA.

This Coffee Consistency Problem is so huge that brute-force maths can't solve it. The mathematics, chemistry and physics of moving some 2,000+ chemicals into your cup of espresso are truly complex. One of the consortium's members, the mathematician Jamie M. Foster, said that to solve it, 'you would need more computing power than Google has …'

Luckily for us coffee-lovers, there is a solution. Surprisingly, this solution comes from the field of renewable energy.

Part of our renewable-energy research looks at energy storage in batteries. For the last few decades, engineers and scientists have been trying to fully understand how lithium ions physically migrate through the electrodes of a lithium battery.

By a lovely coincidence, these lithium-movement equations provided the perfect mathematical shortcuts for the coffee scientists to use.

Even with this lucky break, the Coffee Consistency Problem still took four years to solve. Because their research involved brewing thousands of espressos, at least the team had access to plenty of coffee while doing the maths.

IS FINER BETTER?

Now, it seems 'obvious' that the more finely you grind the coffee, the more chance you have of extracting the full range of 'hidden' coffee chemicals – giving you more coffee goodness and a wonderful taste. After all, a finer grind gives you more surface area for the pressurised hot water to draw the chemicals from.

That's correct, but it leads to other problems (especially clumping). I always imagined that when you made coffee, the very hot water would flow from top to bottom of the coffee basket, and would diffuse through nice and evenly, visiting *all* the grains of

'Extraction Yield' does not equal 'taste'

One factor that we can measure quite easily is the Extraction Yield (EY). This is how much of the mass of the dried coffee beans ends up in the espresso. Yes, the ground coffee that is left behind in the basket has lost some mass – i.e. the chemicals that exited the grains to mix in with the pressurised hot water.

Over many decades of making, measuring and tasting coffee, we have arrived at the conclusion that a good espresso should have an EY in the narrow range of 17–23%.

If you get more than 23% of the chemicals in the coffee beans leaching into the espresso, it tastes bitter; if you get less than 17%, it tastes sour.

But this factor (Extraction Yield) definitely does not give us all the information we want. Two coffees can have the same EY but taste completely different. The EY tells you nothing about the deep subtleties of how wonderful a coffee tastes. That evaluation happens in the taste buds of the barista and, ultimately, of you, the drinker.

Which coffee is strongest?

The percentage of coffee chemicals in the cup of coffee you drink can vary enormously, depending on how you extract these chemicals. Water that is hotter and pressurised (as in espresso) can pull out more chemicals, giving you a stronger coffee than if you simply pour hot water over ground coffee.

The coffee liquid that is espresso contains about 7–12% chemicals (by weight) from the ground coffee, while the liquid that is filter coffee contains only 1% chemicals.

coffee. The result would be a democratic and fair removal of coffee goodness from each grain, which would give you the most flavour.

But no. 'Random clumping' of coffee grains in the basket was the consortium's surprise finding.

If the coffee beans are too fine, the pressurised hot water being pushed through will randomly coagulate some of the ground coffee into an almost solid lump, and the very fine loose particles of ground coffee will wedge in between the bigger particles. But water takes the path of least resistance, and in going around that solid lump of coagulated powder, the water will open up channels.

So now the hot water is gushing through the channels that it pushed open – and is bypassing a lot of the coffee grains. This is far from perfect.

On one hand, you get very little coffee goodness from those grains that coagulated into a lump. And on the other hand, you take too much of those 2,000+ chemicals from the grains lining the channels.

Random clumping is the enemy of consistency. The fine grind and the high water pressure 'choke' the espresso machine.

The consortium thought deeply and experimented. Finally, they arrived at their Better Coffee Conclusion, which made these three recommendations: use less coffee, do less grinding, and use lower water pressure, which gives a faster shot time.

As a bonus, it's cheaper to use less coffee. One coffee shop in Oregon was able to save 13 cents per cup, which added up to US$3,620 per year. Spread across the USA, that saving is about US$1 billion per year.

To make a more perfect and consistent espresso, we should use three quarters as much coffee (15 grams instead of 20) and grind the coffee beans less finely. In addition, run the water at a lower pressure (6 atmospheres, not 9), and extract the espresso for a shorter time (15 seconds).

It turns out that coarser grounds can give a consistent 23% Extraction Yield of just one flavour. But finer grounds, while producing the same Extraction Yield of 23%, give a variety of different flavours, possibly including unpleasant under- or over-extracted flavours.

Of course, the proof is in the drinking. Already, some baristas are saying that this new regime produces espresso that is better tasting and more consistent from one cup to the next.

It's not exactly an independent double-blind trial, but after 600 years of taking a shot in the dark, we might be on track for the perfect espresso every time.

POINT BREAK

THE BUTTERFLY EFFECT ON COFFEE

The random clumping of ground coffee in an espresso machine is another example of the 'Butterfly Effect'.

In 1961, the meteorologist and mathematician Edward Norton Lorenz accidentally rediscovered Chaos Theory (popularly called the Butterfly Effect). Chaos Theory deals with systems (weather, coffee in an espresso basket, etc) where a tiny change to the conditions at the start leads to a profound difference in the final result.

He was running a computer program that tried to predict the weather. He found that if he changed just one data input (say, the air pressure or the temperature) by the tiniest amount (say, 0.01%), the resulting weather prediction a few days later would be completely different.

This so-called Butterfly Effect is actually 'sensitivity to initial conditions', if you want to be technical.

Start with a butterfly flapping its wings in the Amazon rainforest. Supposedly that microscopic change in the air ripples through the weather systems. As a result, a fortnight later, a hurricane changes course. Instead of destroying a US city, it stays entirely at sea.

The Butterfly Effect also happens with coffee grains in an espresso basket. Suppose that the grains in one section are packed just a little more tightly – and by 'a little', I mean by 0.01%. There's no way a human can feel that difference.

That tightness of packing is the tiny change in 'initial conditions'. But it has a flow-on effect. That section of coffee grains could clump into an almost-solid blob, and the hot water would simply flow around it – and not extract any chemicals from it.

The end result is not a hurricane, but a slightly less-than-ideal cup of coffee.

DEAD FISH SWIM

In Mafia movies, 'He is sleeping with the fishes' means only one thing – and it's not good. In gangsta talk, the expression tells you somebody has become dead, and it sure wasn't from natural causes.

But while dead men don't walk, dead fish *can* swim – and we accidentally proved it at home.

A SAD FISHY TALE

When our daughter was little, she wanted fish in a fish tank, so of course, we got them. Everything went swimmingly for a while – happy fish and happy kid.

Until, one fateful day, a visitor said, 'You know one of your fish is dead, right?' We were astonished. After all, it looked like the fish was still swimming. There he was, right next to the filter outlet, where the cleaned water was being pumped into the tank. (And that location was the big clue, which we didn't understand at the time.) Our fish looked like he was swimming normally – up to the filter, getting pushed back, swimming up to the filter again, and repeating this continuously in a loop.

But then we saw the Big Picture. That was all he was doing. He wasn't swimming anywhere else in the tank.

Our visitor was correct – our beloved fish had been dead for a day or two. We hadn't noticed, because he actually looked like he was swimming properly.

We didn't have a Wonder Fish, we had a dead fish. What was going on?

DEAD FISH CAN SWIM

Our inability to recognise that our little fish was a goner was because we had failed to read the 2006 academic paper in the *Journal of Fluid Mechanics* titled 'Passive propulsion in vortex wakes'. (Another problem was that we hadn't been paying our fish enough attention.)

This paper describes how a dead fish can indeed move upstream – against the oncoming water. The very first sentence starkly laid it

Sleeping with the fishes

This expression goes back much further than *The Godfather* movies.

The author Edmund Spencer, used it in 1836, in *Sketches of Germany and the Germans*: ' … if he repeated his visit, they would send him to sleep with the fishes.'

But this phrase goes back much further. Homer used something similar in *The Iliad*, 2,800 years ago: 'Make your bed with the fishes now …'

out for us: 'A dead fish is propelled upstream when its flexible body resonates with oncoming vortices ...' Yep, 'dead fish' were the second and third words.

The authors write that 'anecdotal stories from whalers tell of dead whales coasting at approximately one knot for long periods of time ...'

Somehow a (dead) fish can extract energy out of the water that is flowing directly towards it.

PHYSICS OF DEAD FISH

When a fish is alive and gliding through the water, it usually flaps its tail from side to side. Part of the swimming movement of the tail involves pushing water backwards, away from its body.

But how does flowing water make a dead fish swim? As a first step, let's assume that the water flows, at a constant velocity, directly towards the dead fish. When it hits the head of the fish, it creates turbulence – such as little whirlpools. Further downstream, oscillating spirals of water flow down the side of the fish. These force the body to curve – first one way, and then back again.

If the timing is right, one of these spirals of water can hit the front of the tail when the tail happens to be bent sideways – say, at an angle of 45°. The fast-moving spinning water forces the tail backwards, which pushes against the water behind it. Newton's Third Law of Motion kicks in – 'For every action, there is an equal and opposite reaction.' So if the water behind the tail is pushed backwards, then the fish is pushed forwards.

Now, this is beginning to seem like perpetual motion. Where is the dead fish getting the energy from? Well, it has to be from an external force. In our sad case at home, the energy was coming from the electric pump, which was pushing water towards the fish.

There's no magic, or physiology, for that matter. All the unexpected movement of the dead fish comes entirely from the energy in the moving water.

PHYSICS OF LIVE FISH

Of course, evolution being what it is, sea creatures have evolved to take advantage of the energy boost that comes from the water moving around them – both with them and against them.

Most fish have a 'lateral line'. This is a row of sensors that detect (among other things) movement, vibration and pressure. Fish use the lateral line to find fleeing prey, or avoid an incoming shark. They almost certainly also use it to make life more cruisy – such as adjusting their body movements for greater efficiency of energy use.

A whale can get up to 25% of the power it needs to swim forward when it's travelling against a head sea by adjusting its body this way and that. (A head sea is when the waves come directly head on.) That 25% improvement in energy use is pretty impressive. The whale gets only a bit more power (another 10% to bring the saving to 35% in total) when the waves are pushing it along from behind. Again, that sounds like free energy, but it's not. The energy comes from the Sun, which is the main driver of weather on the planet (yes, the Sun powers the weather, which in turn powers the winds, which in turn power the waves). This is because each square metre of sunlight carries about 1.5 kilowatts of power, before it hits the atmosphere. But regardless of where the waves are coming from, the whale can use them for a free ride.

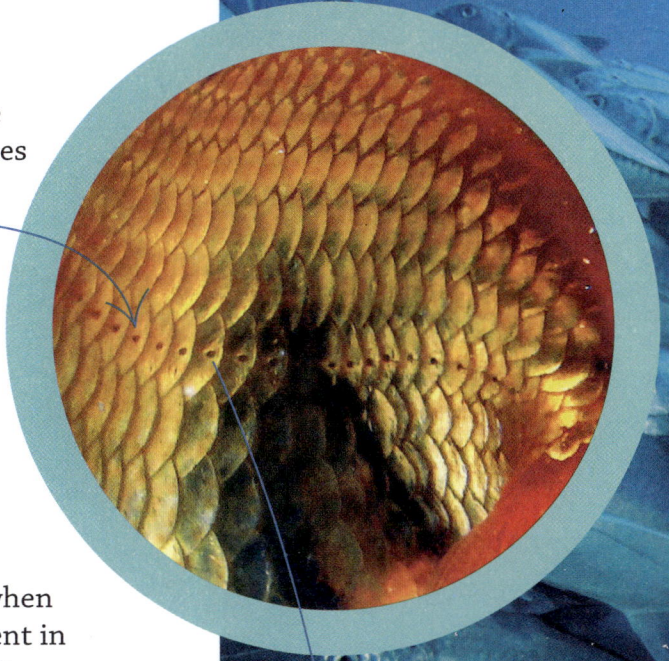

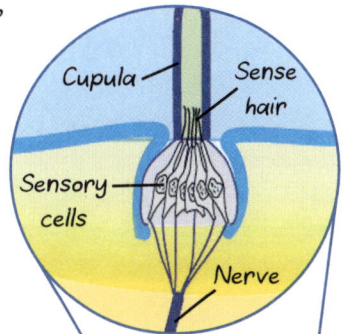

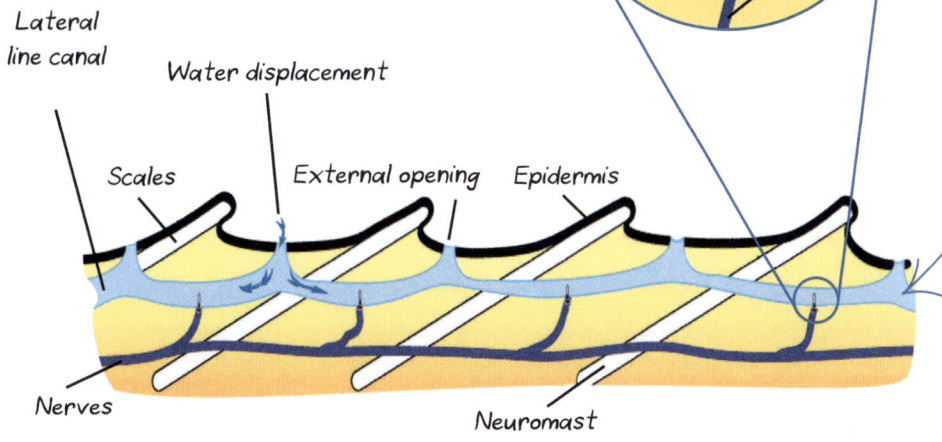

Cupula

Sense hair

Sensory cells

Nerve

Lateral line canal

Water displacement

Scales

External opening

Epidermis

Nerves

Neuromast

15

Dolphins that gracefully skim the bow wave of a moving ship are also getting a free ride. In this case, the diesel engines of the ship have to work just a tiny bit harder so the dolphins can benefit. Fish swimming in schools use similar techniques to get 'free' energy.

Let's look at the first row of fish in the school. Those fish use some of the energy in the water flowing towards them to push themselves forward. So they benefit, and the water slows down. No wonder fish go to school – all that 'free' energy!

FLOW-ONS?

I can see a few benefits flowing from this.

First, we could work out more creative ways to extract energy from moving water. Sure, we already use turbines, but there might be advantages to using different technologies, such as vertical bendy slabs that gently wave in the passing waves to make electricity.

Second, perhaps some sports scientists can use this knowledge to work out a way to give Olympic swimmers a secret advantage.

Maybe we've been wrong in urging swimmers to swim like (live) fish? Perhaps the secret is to copy the dead ones.

DEAD FISH SWIM

Q+A

Do fish feel pain?

Yes, every animal has to know if something bad is happening to its body. So sea creatures, from tiny fish to blue whales, have specialised cells that respond to different types of pain (sharp or dull pressure, vibration, heat, cold, etc) and send electrical signals via nerve pathways to specialised areas in the brain.

If you inflict pain on a fish (e.g. by shoving a big metal hook into its face), it will respond by making certain characteristic movements. But if you give the fish an opiate (e.g. morphine) and then shove the hook into its mouth, then it won't move in that very specific and characteristic way. Yup, fish feel pain, as humans do, and just as in humans this pain can be blocked by morphine.

So to reduce the amount of pain a fish feels, as soon as you catch it place it into ice-cold water which will, at least, anaesthetise it and send it to sleep.

17

SELF-REPAIRING LUNGS

Most of us know that smoking is bad for your health. But does giving up smoking improve things?

The good news is yes. We've known that for a while. We just didn't know why. But recently, we have begun to realise that stem cells seem to play a part.

MYSTERIES

Mark Twain said, 'Giving up smoking is the easiest thing in the world. I know because I've done it thousands of times.' Giving up smoking for good is indeed hard work – which is no surprise because nicotine is a very addictive drug. The best way to stop smoking (nicotine patches, cold turkey, hypnotherapy, etc) is still a bit of a mystery.

And, when it comes to ex-smokers, there are mysteries to do with their healing.

First, very shortly after a person gives up smoking, their risks of getting a lung cancer begin to drop.

But why does this happen, and so quickly? Surely the damage that has been done is permanent. How can cells repair themselves? After all, tobacco smoke brings on lung cancers by increasing mutations in lung cells. And unfortunately, mutations don't spontaneously fix themselves.

Second, if we compare two ex-smokers who have, in total, smoked the same number of cigarettes over their lives but gave up at different times, we can see two different outcomes.

Person A really loved smoking, and smoked lots of cigarettes every day. For some reason they stopped after 20 years of smoking. Imagine that it's now ten years since they stopped smoking.

Person B smoked fewer ciggies each day. It took them 29 years to smoke the same number of cigarettes, in total, as Person A. Last year, they also stopped smoking, and now it's one year later.

In all other ways, they match up. They're the same age, have smoked as many cigarettes and are otherwise pretty similar.

But the bare statistics tell a different story.

Person A, who gave up ten years ago, has a lower risk of cancer than Person B, who gave up only one year ago. The benefits of giving up smoking start as soon as somebody stops, and continue to increase with time. And that's true no matter how many ciggies they smoked.

Tobacco kills

We know that, according to the World Health Organization, over 1.1 billion people worldwide smoke cigarettes.

Each year about 1.8 million smokers die from one of the several different types of lung cancer. We also know that lung cancers take decades to appear, and smokers are some 30 times more likely than non-smokers to get a lung cancer.

And finally, lung cancers kill more people than any other cancer.

Air comes in via your trachea, which then splits into successively smaller pipes. Bronchi are some of the pipes along the way. There are some 22 sets or levels of splitting through about 2,400 km of airways until the air finally arrives at the 400 million alveoli, where oxygen is taken up by the blood.

But shouldn't their health risks be the same, if overall they smoked the same number of cigarettes?

It's these kinds of questions that intrigue the scientists who work in lung cancer.

'SLEEPING' UNDAMAGED STEM CELLS

We are moderately sure that there is a reservoir of near-normal stem cells somewhere in the lungs. But we don't know exactly where they are.

We think these stem cells don't do a lot of dividing, because their telomeres are quite long. They seem to spend a lot of time resting – or, at least, not being active. And then, when the smoker gives up cigarettes, these stem cells come charging out of their hiding place to replace cells that were damaged by years of exposure to the cancer-causing chemicals in burnt tobacco. The new and exciting finding is that the damaged cells don't get repaired – they get completely replaced!

Bronchi, Bronchial Tree and Lungs

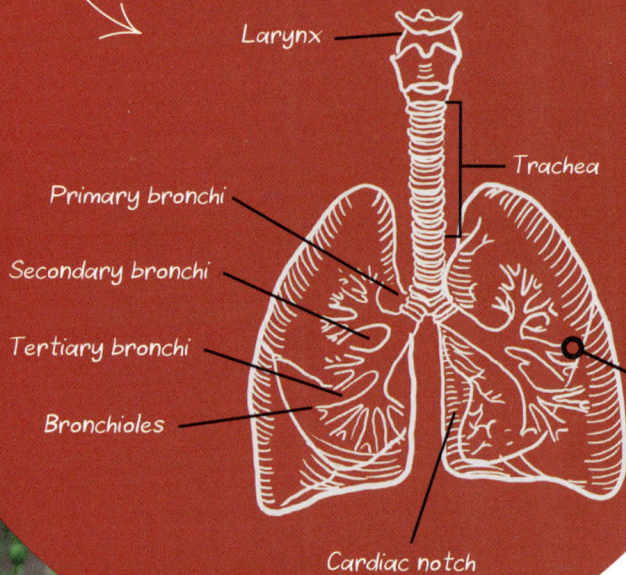

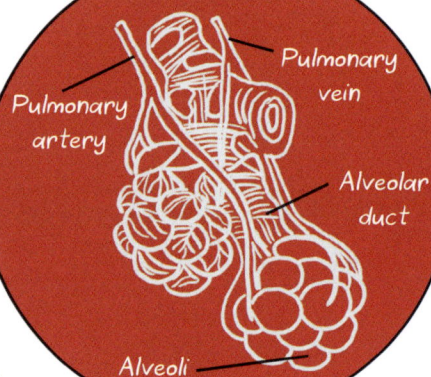

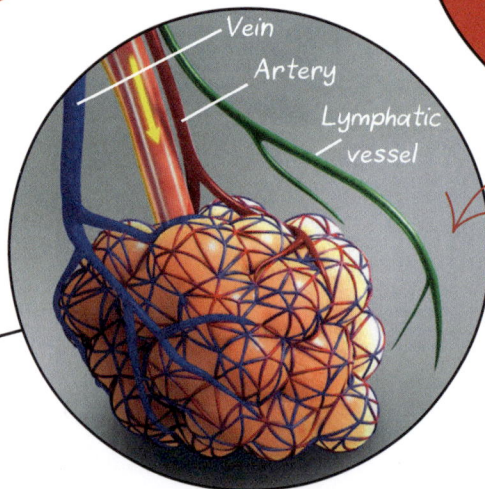

Larynx

Trachea

Primary bronchi

Secondary bronchi

Tertiary bronchi

Bronchioles

Cardiac notch

Pulmonary artery

Pulmonary vein

Alveolar duct

Alveoli

Vein

Artery

Lymphatic vessel

The alveoli are where gas exchange happens. Oxygen leaves the lungs to enter the blood. At the same time, carbon dioxide goes in the other direction. Alveoli have a total surface area of 70–140 sq metres (nearly half a tennis court).

POINT BREAK

TELOMERES & AGEING

Telomeres (the little bits on the ends of your chromosomes) have a similar job to aglets (the little plastic or metal bits on the ends of your shoelaces).

Aglets stop the ends of your shoelaces from fraying and getting damaged. Once an aglet falls off, the end of the shoelace frays rapidly. Very soon, it's quite hard to thread the shoelace through the holes in your shoes.

Cells usually divide 50–70 times before they die. Inside the cell, the DNA is packaged into chromosomes. Telomeres protect the ends of your chromosomes. But each time your cells divide, the telomeres get shorter. So, long telomeres tell you that the cell hasn't done many divisions, and is quite young. And short telomeres indicate the cell has been through many divisions.

Elizabeth Blackburn, a biochemist from Tasmania, made the link between telomere length and cell health/age. She won a Nobel Prize in 2009 for her discoveries in this area.

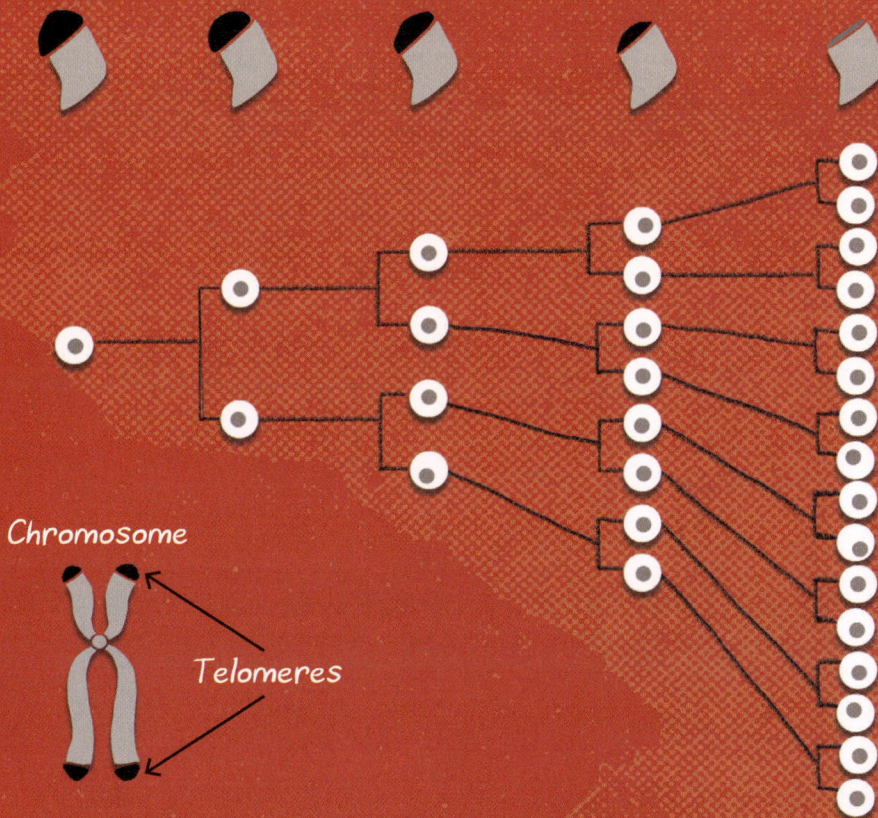

Chromosome

Telomeres

Cells usually divide 50–70 times before they die. Inside a cell, the DNA is packaged into chromosomes. Each time the cell divides, the shorter the telomeres on the ends of the chromosome become.

Cell

Nucleus   Chromosome   DNA

5' end

Adenine   Thymine

Phosphate-
deoxyribose
backbone

3' end   Guanine   Cytosine

5' end

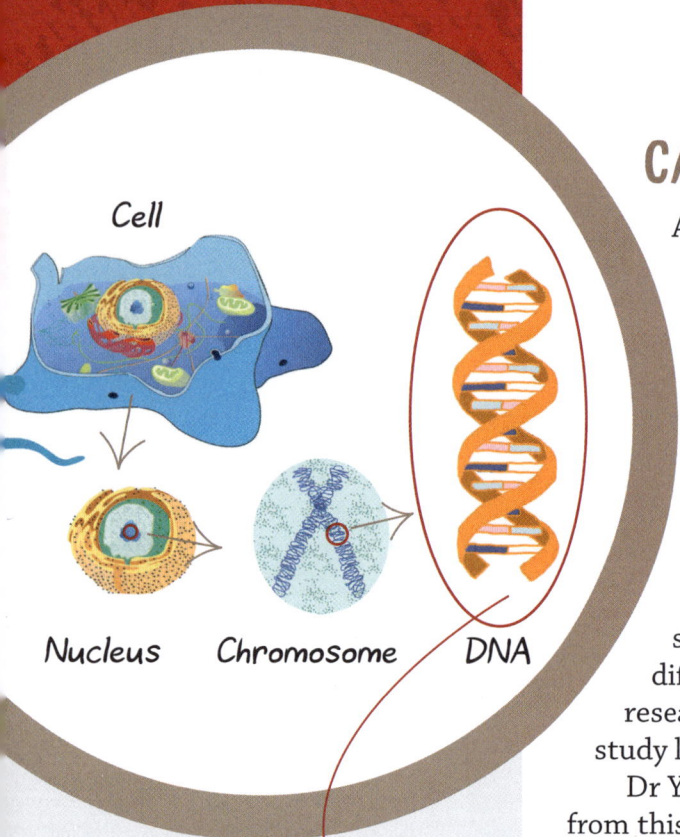

CANCEROUS CELLS

A study by Dr Kenichi Yoshida and colleagues looked at the bronchial epithelial cells in four groups of people. (The bronchi are pipes that carry air deep into the lungs all the way down to the alveoli, where gases get swapped over – oxygen enters the blood and carbon dioxide leaves it. The epithelial cells line the bronchi, forming a protective barrier between the air and the body.)

The four groups were: three young children who had never smoked; four adults who had never smoked; six people who had smoked but had quit; and, finally, three people who were still smoking cigarettes. (So this small study did not check people who had given up smoking at different times in the past. At least this is a signal to other researchers in the field to follow along this pathway, but to study larger numbers of people.)

Dr Yoshida's study looked at 632 healthy lung cells gathered from this very small sample of just 16 people. The scientists were looking for mutations in the DNA of bronchial cells.

Smokers had more mutations than non-smokers in their DNA. And smokers had a different pattern of mutation from non-smokers. Their most common mutation was guanine-to-thymine (in other words, in their DNA, the nucleotide called 'guanine' would get turned into 'thymine'). We are still trying to understand the significance of this finding.

As an aside, cancers of the lung carry more mutations in their DNA than practically any other cancer. (Which is another reason that it's such a shame that smoking is so addictive!)

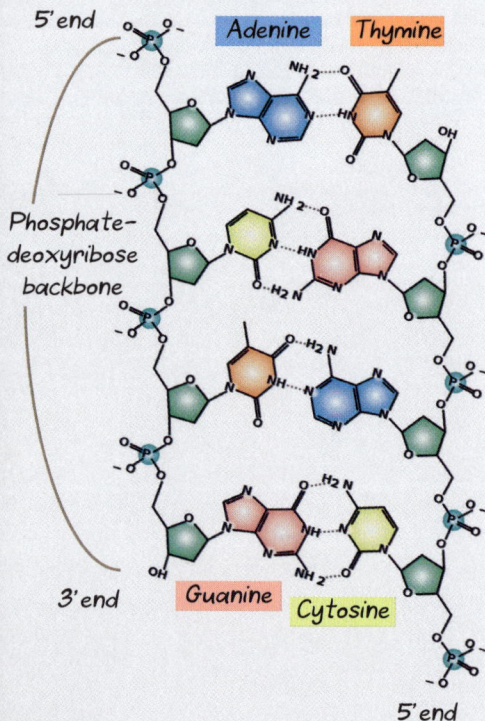

THE RESULTS – GOOD AND BAD

As you get older, the DNA of each cell in your body acquires more mutations. Each extra year of life adds about 22 mutations to each cell – even if you don't smoke. Bummer, let's never get older! (Luckily, most mutations don't lead to cancer.)

So, being 60 years old means you carry 1,320 mutations per cell. The Yoshida study found that ex-smokers had an additional 2,330 mutations on top of this. Worst of all were current smokers, with an additional 5,300 mutations.

Middle-aged adults who'd never smoked had 4–14% of their bronchial cells carrying mutations that could lead to a cancer. But current smokers were worse off – 25% of their bronchial cells had mutations.

One odd thing was the variability in the number of mutations. The non-smokers had quite a small variability from cell to cell. But in the lung cells from a single smoker, the number of mutations in any given cell could vary by a factor of ten, say, from 1,000 to 10,000.

And finally, in ex-smokers, the number of mutations actually dropped compared with those of the smokers! Yes, some unknown healing mechanism had kicked in. In fact, 20–50% of the cells of ex-smokers looked just like the cells of somebody who had never smoked.

The scientists are guessing that these healthy cells arose from stem cells. The ex-smokers had up to four times as many of these healthy cells as current smokers.

These 'undamaged' cells hold the secret. They had really long telomeres, implying that they had arisen, comparatively recently, from (here comes the guesswork) 'proposed previously dormant (quiescent) stem cells'. Always searching for the truth, good scientists point out that they haven't actually found these 'sleeping beauty' cells yet. So they write, 'Whether such cells exist in human lungs is unknown.'

And there are so many more questions. How did the stem cells (if they really are the repair/replacement cells) survive undamaged, despite a few decades of exposure to tobacco smoke, which carries at least 70 known carcinogens? And why did only some, not all, of the damaged cells with mutations get replaced?

HAPPY-ISH ENDING

So, we still don't know the location of the sleeping stem cells (or even if they exist). We also don't know how they avoided the high rates of mutation that happened with neighbouring cells.

And finally, we don't know the pathway by which this population of untouched stem cells expands and migrates to where it's needed, when a person gives up smoking.

What we do know is that stopping smoking may awaken cells that have not been damaged by sucking on 'cancer sticks'. And we also know that the sooner you stop smoking, the better off you'll be. But no matter when you give up, your health will improve.

Stem cells have special powers

Stem cells are special cells that can do two things.

Their first 'special power' is that they can self-renew. In other words, they can make more stem cells by cell division. This means that you can maintain an immortal 'reservoir' of stem cells until death. Unfortunately, at the moment, our knowledge of stem cells is not great enough to make our whole bodies immortal, yet …

Their second 'special power' is that they can differentiate. This means they can turn into many different types of specialised cells (e.g. skin, liver, heart, lung).

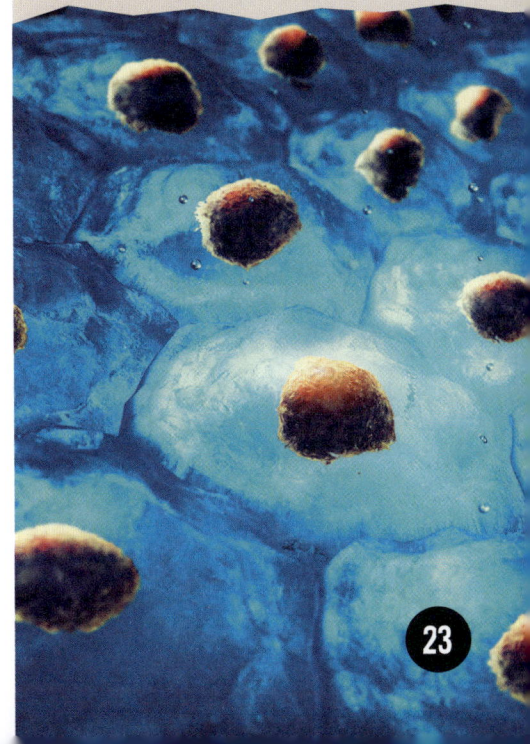

EASTER & EQUINOX

Like most Australians, I enjoy a public holiday. Most public holidays are cleverly and conveniently organised to fall on the same date each year, such as New Year's Day, Christmas Day and Anzac Day. Which sure does save confusion.

But Easter happens anywhere between 22 March and 25 April – why the big variation in dates?

The answer is a mixture of religion and astronomy.

EASTER TIMING 101

Easter has been a symbol of new life and fertility for millennia. The early Christians simply took over that ancient festival and rebadged it as a celebration of the Resurrection of Christ from his tomb, three days after he died.

But in the early days of Christianity, the exact time to celebrate Easter varied across the world. In 325 CE, the First Council of Nicaea laid down two different rules for working out the timing of Easter. They neglected to give any details for how to work out the exact date. In fact, they did not even specify that Easter had to be on a Sunday. Furthermore, their rules were based on the old Julian calendar, which carried within itself a few astronomical errors – so it began to drift out of sync with the astronomical year (by 1582, it was ten days wrong). As you can imagine, all this led to a few problems.

Several centuries later, in 725 CE, the English Benedictine monk known as St Bede, or the Venerable Bede, laid down the law on when to celebrate Easter. He declared, 'The Sunday following the full Moon which falls on, or after, the Equinox, will give the lawful Easter.' (He meant the Spring Equinox, which happens in March in the Northern Hemisphere, but let's talk about equinoxes later.)

Pope Gregory XIII reaffirmed this rule in 1582. Yup, Pope Gregory is the dude responsible for the Gregorian calendar, which we still use today.

On average, with these rules, the date of Easter slips back about eight days each year, and then it jumps forward again.

An egg-cellent history

Around the end of March each year, the northern hemisphere wakes up from its long Winter nap. Spring is coming – and so is Easter.

In many older European religions, the end of Winter was celebrated with the arrival of a special Spring Goddess. The Scandinavians had 'Ostra', the Anglo-Saxons 'Eostre', and the Germans 'Eastre'. (You can see they all sound like 'Easter'.)

Eggs have been a feature of Spring/Easter celebrations right up to the present. That's because eggs are a magnificent symbol of new life. For centuries, the Christian religion forbade eating eggs during Lent (the season of fasting). Being able to eat eggs again at Easter was a special treat.

Rabbit-like animals, revered as symbols of fertility and the new life that came with Spring, have long been connected with Easter, too. The Easter Bunny himself came to us from the Germans. He first appeared in the 1600s, delivering coloured eggs on Easter Sunday to delighted children after the hard slog of Winter.

The first chocolate Easter eggs appeared in Europe in the early 1800s, after the Dutch invented a press to separate cocoa butter from the cocoa bean in 1828.

But the ultimate Easter eggs are the amazing bejewelled creations of Peter Carl Fabergé, fashioned between 1885 and 1917 for the Russian Royal Family.

ASTRONOMY & EASTER

Let's break down the Venerable Bede's instructions: Equinox + Full Moon + Sunday = Easter Sunday!

'Sunday' is the easiest part to understand – it's just the day in the week between Saturday and Monday.

'Full Moon' is when the Moon is completely lit up by the Sun. Sometimes it's a bit hard just by looking at the Moon to tell if it is full – or just before or just after. However, a few days of watching the Moon rise will sort that out. One good hint is that if the Moon rises in the east at sunset (not an hour before or an hour after), that's a full Moon. (The outer circle of moons in the diagram is the observer's view from Earth.)

So next on our list is an 'equinox'. What's that?

The Latin roots give you a hint – 'aequus' for 'equal', and 'nox' for 'night'. So the Equinox is that day when there are equal hours of daylight and night-time. There are two equinoxes each year – one in Spring and one in Autumn.

On what exact day do these equinoxes happen? Well, it's related to the Terminator – which in this case has nothing to do with Arnold Schwarzenegger acting as a killer robot from the future.

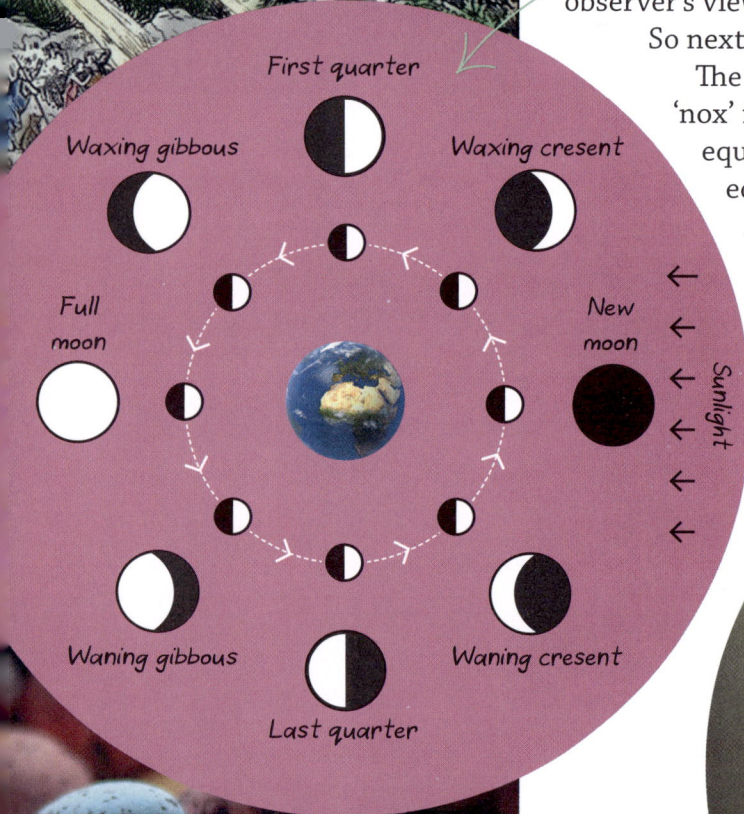

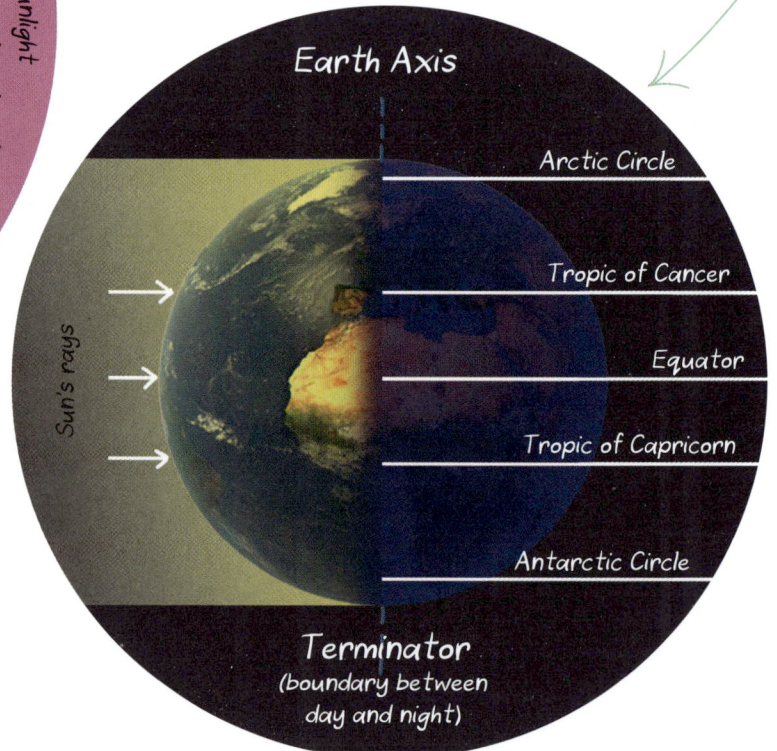

First quarter

Waxing gibbous

Waxing cresent

Full moon

New moon

Sunlight

Waning gibbous

Waning cresent

Last quarter

Earth Axis

Arctic Circle

Tropic of Cancer

Equator

Tropic of Capricorn

Antarctic Circle

Sun's rays

Terminator
(boundary between day and night)

In astronomy, the Solar Terminator is the line between day and night. The Equinox happens when the Terminator crosses both the North and South Poles simultaneously. (And yes, this means that each point on the surface of our planet gets almost-equal minutes of sunlight and darkness.) This date happens in mid-Spring and in mid-Autumn (around 20 March and 22/23 September).

A NASA photo of the Solar Terminator, March 2019

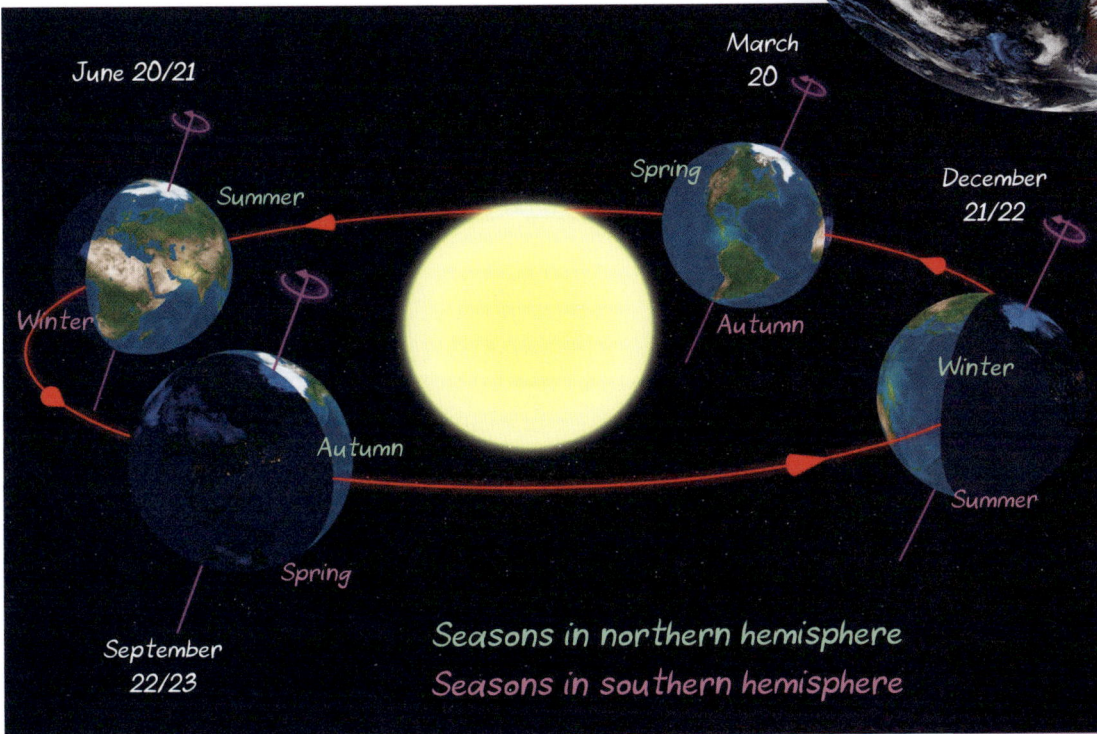

Earth's seasons

June 20/21

Summer

Winter

Spring

March 20

December 21/22

Autumn

Winter

Autumn

Summer

Spring

September 22/23

Seasons in northern hemisphere
Seasons in southern hemisphere

Now, let's get back to the Venerable Bede's three-step rule – and put it all together.

First, we wait for the date of the March Equinox, which is when there are pretty equal hours of daylight and darkness.

Second, we wait for the next full Moon, which can be any time from that same day to 28 days after the Equinox.

Third, we wait for the next Sunday to come around, which can be at any time from one to seven days away.

And that's Easter Sunday.

It's certainly not a simple rule (thanks heaps, St Bede).

Fixing Easter?

A movement began in the early 20th century to simplify things and change Easter to a fixed date. One rather popular date was the second Sunday in April.

The Easter Act 1928 in the UK proclaimed legislation for a fixed date of Easter, but the Act was never implemented. The Second Vatican Council in 1963 said it was fine with fixing a date, so long as all the Christian churches could agree.

So far? No consensus whatsoever. Nothing, nada, zip, zilch, zero. We still have a floating Easter.

RULES ARE MEANT TO BE BROKEN ...

But wait, it gets worse. On very rare occasions, you don't even follow the rule at all.

An example was the March Equinox of 2019. The Solar Terminator crossed the North and South Poles simultaneously at 9.58 pm Universal Time on Wednesday 20 March. Pretty straightforward – that's the exact time of the Equinox (the first step).

The full Moon happened a few hours later, which happened to be the next day, at 1.43 am Universal Time on Thursday 21 March (the second step).

Obviously, the next Sunday, 24 March, should have been Easter Sunday (the third step). But it wasn't.

In 2019, Easter Sunday was celebrated on 21 April – nearly a month later!

What's. Going. On?

RELIGION VS SCIENCE – WHO WINS?

The astronomers are quite comfortable that the March Equinox is not a fixed date, but can happen on 19, 20 or 21 March. (There's a whole bunch of complicated astronomy reasons why. Let's ignore the tricky details.)

But regardless of what the astronomers reckon, the religious definition is that the Equinox happens on 21 March, as decreed by Pope Gregory XIII – and for religious celebrations, the religious definition always takes priority.

So in 2019, the Equinox was on 20 March. The full Moon happened on 21 March. The next Sunday was 24 March. But was it Easter-egg time? Nope.

Remember that Pope Gregory had defined the Equinox to be on 21 March. The next full Moon *after* that was 19 April, and the next Sunday was 21 April. So *that's* when the chocolate came out in 2019.

Who knew that working out when Easter Sunday should fall would be such a complex business? Maybe we should excuse the supermarkets for hedging their bets with chocolate bunnies on the shelves from February ...

But one thing is for sure. When Easter is just around the corner, I'll be in my Easter Bonnet waiting for the Easter Hat Parade and the Easter Bunny, and the infinite chocolate ...

EQUINOX IS NOT 'EQUI'

There's another problem with the Equinox – it's not 'equi'.

You would think that on the 24-hour day of the Equinox, there would be absolutely equal minutes of daylight and night-time.

But no! Check out the sunrise and sunset times on 20 March 2019 (which the astronomers labelled as the Equinox). There were more minutes of sunlight than of darkness (6 minutes more if you were on the Equator, and 26 minutes more at Longyearbyen, in Svalbard, inside the Arctic Circle).

Why is it not exactly 12 hours? The answers are to do with 'bent light' and a 'poor definition of sunrise/sunset'.

The first part of the answer is that our planet has about 5,000 trillion tonnes of atmosphere. The atmosphere bends the light of the Sun. (You've heard of mirages. And you've probably seen 'water' glistening on a hot two-lane blacktop road, halfway between you and the horizon? That's not really water – that's the light from the sky being bent by the hot air immediately above the road, and instead of landing on the road, it gets curved upwards to land in your retina.)

Suppose you've found a nice picnic spot, and you and your friends are there at sunset. Then comes that lovely moment when the whole Sun is red, *and* the lowest part of the Sun is just kissing the horizon.

But the atmosphere is tricking you. This image of the Sun is an optical illusion.

In reality, the whole of the Sun at this moment is already totally below the horizon. If there were no atmosphere, we would see the Sun do its setting-below-the-horizon in real time. But there *is* an atmosphere, which bends the light, bringing it above the horizon to land in your eyes.

Now, here comes a very nice coincidence – which always amazes me. (I'm easily amazed.) The atmosphere bends the light of the Sun by half a degree, which also happens to be the diameter of the Sun. So when you think you see the *bottom* of the Sun just kissing the horizon on its way down, in reality the *top* of the Sun has just vanished below the horizon.

The second part of the answer is that the Sun is not a point with no size – it's a disc about half a degree across.

This would not be a problem if we defined sunrise and sunset as when the centre of the Sun crossed the horizon.

But no. Somebody (and we don't know who) in the distant past chose a different definition. Sunrise and sunset happen when the tiniest part of the Sun is just visible. So this adds a few extra minutes to the day, even on the Equinox.

You

Apparent but fake position of the Sun

Actual real position of the Sun

SPIDERS CAN FLY

The spider on the right is in the 'tip-toe' position, about to launch into flight.

I love a good surprise, and this is a great one.

What kind of wingless creature can fly for over a thousand kilometres non-stop, at altitudes higher than four and a half kilometres?

Give up?
Amazingly, it's a spider.

IT'S RAINING RED SPIDERS

Nearly two centuries ago, the English naturalist Charles Darwin was equally amazed. He was partway through his famous voyage (1831–36) around the world on the HMS *Beagle*. The ship was way out at sea in the Atlantic Ocean, about 100 km off the coast of Argentina, a long distance from the sight of land.

On 31 October 1832, a hot and relatively calm and clear day, their ship was suddenly inundated by vast numbers of red spiders. These red spiders literally dropped out of the sky, just like in a B-grade horror movie.

The spiders came in two sizes – the smaller ones were around 2–3 mm across, while the larger spiders were about 7 mm across. Darwin thought that they were probably juveniles of two different species. They all had silk threads (or gossamer) hanging from them. Darwin wrote that there were so many that 'all the ropes were coated and fringed with gossamer web'.

He watched them keenly from when they invaded his ship until they later took off to continue their journey. He noticed there were quite different launching behaviours and patterns, depending on the size of the spider.

Darwin recorded that the small spider would 'elevate its abdomen, send forth a thread, and then sail away horizontally, but with a rapidity which was quite unaccountable'.

But then, being a skilled observer, he noticed that the larger spiders did something quite different with their fine threads of spiderweb. A larger spider would 'dart forth four or five threads from its spinners ... in undulations like films of silk blown in the wind. They were more than a yard in length, and diverged in an ascending direction from the orifices. The spider then suddenly let go of its hold ... and was quite quickly borne out of sight ...'

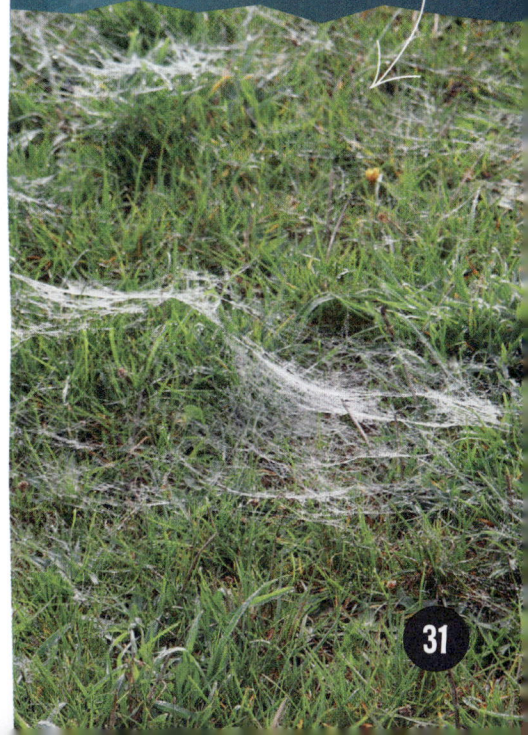

Threads of spider silk left after a mass spider ballooning.

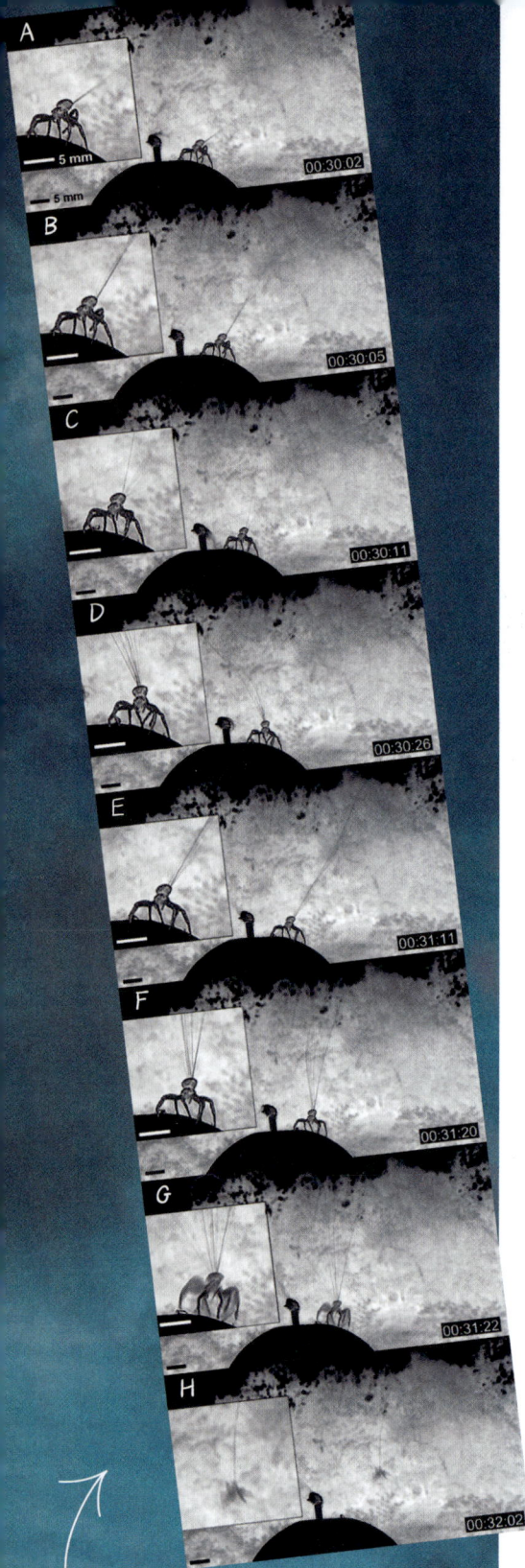

(A, B) Initial phase of spinning ballooning lines; (C, D, E, F) Wind causes unsteady fluttering of a bundle of ballooning lines; (G) Moment of take-off; (H) Airborne state of a ballooning spider.

Notice two different things about the threads coming out of the 'spinners' or 'spinnerets' of the larger spiders – they *lifted*, and they *separated* or *spread out*.

Now, Darwin thought that because the day was hot, there might have been very faint rising thermal breezes, which could have *lifted* the silk threads as they came out of the spinnerets. That's kind of obvious – I might have been able to guess that much.

But his next thoughts were very clever indeed. He guessed that the *spreading apart* of the spiderweb threads was caused by some kind of electrostatic repulsion.

ELECTRIC SPIDERS

You've probably seen electrostatic repulsion (static electricity) in action. Rub a balloon on a shirt, and then hold it close to the head of somebody with long, flowing hair. The balloon has already picked up an electrical charge by being rubbed. When you hold it close to the scalp, the balloon transfers its charge to the many individual threads of the long hair, which then stand up away from each other. The scalp hairs repel each other, because similar charges repel each other.

Getting back to our spiders, they are actively using the Earth's natural electric field. (Yup, the Earth has a natural electric field, as well as a natural magnetic field.) This field is continuously maintained by the 40,000-or-so thunderstorms that happen each day on our planet. In fine weather, the ground carries a negative charge, while the upper atmosphere carries a positive charge. (This is the opposite of what happens with lightning, when the ground is positive, and the bottom of the cloud is negative.) Amazingly, the scientists still don't fully understand how the different charges (that the ground can carry) are generated.

As the spiders extrude their silk from their spinnerets, the silk picks up a negative charge. And with their charged silk threads, the spiders can then hitch a free ride on the Earth's natural electric field.

The strength of an electric field is usually measured in volts per metre. On a cloudless sunny day, the electrical potential of the atmosphere rises by about 120 volts for each vertical metre. But when it's stormy or foggy, each vertical metre can carry thousands or tens of thousands of volts. So this voltage difference between the ground and the upper atmosphere – say, 50 km up – can be as high as 350,000 volts. That's astonishingly charged!

Now I'm not really sure, but this seems to imply that spiders

UP, UP AND AWAY

Ballooning away from the birth site helps newborn spiders avoid getting eaten by slightly bigger spiders. It also reduces competition for resources (such as food) in an area densely populated by young spiders. The ballooning newborns (aka spiderlings) can survive for up to 25 days without food. However, the death rate can be high – just from some of the hazards they might encounter while in flight.

Ballooning is mostly done by spiderlings, but larger adults also do it. Some spiders up to 14 mm long, and weighing more than 100 mg, have been spotted ballooning. Ballooning happens most often during late Spring and Autumn.

Ballooning is probably how, when a brand-new volcano pops up out of the ocean, one of the first immigrants to arrive is (you guessed it) spiders. There might be nothing for them to eat, but 'luckily' spiders have a tendency to cannibalism.

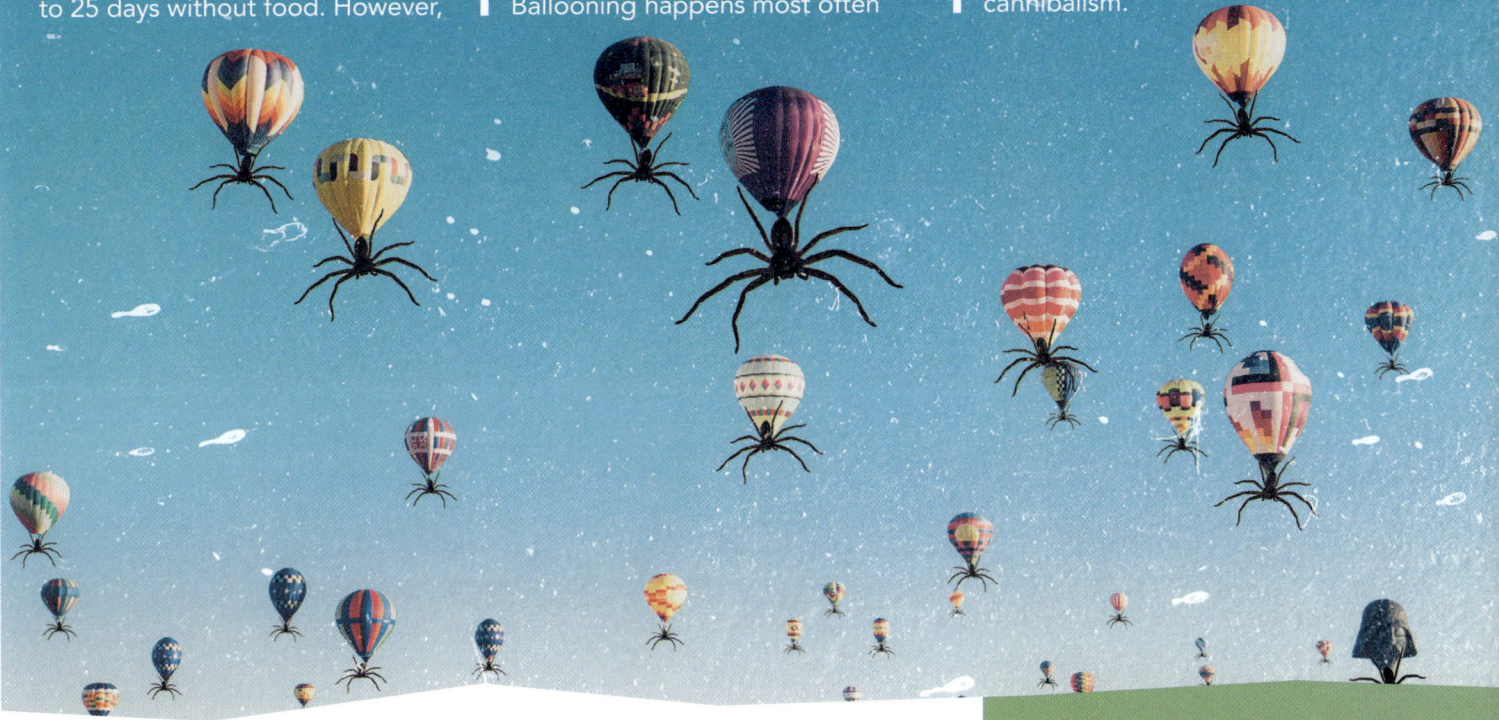

can fly 'better' on stormy or foggy days, because of the extra charge in the atmosphere. But the other hazards associated with flying through a storm might convince the spiders to not fly on those days. After all, only a few surfboard riders can ride the biggest waves – and most surfers avoid them.

CATCH THE WIND

Like all good aviators and pilots, flying – or, technically, ballooning – spiders will test the wind before they fly. They normally try to find a launch site that's a bit higher than the surrounding countryside.

Electric bees?

Bumblebees can detect, and respond to, the electric fields that exist and arise between flowers and themselves.

Honeybees actually use electrical charges to communicate with other honeybees inside their hive.

Charged spider silk

We're not entirely sure how the spider silk picks up an electrical charge as it's manufactured and squeezed out through the spinnerets at the back end of the spider.

Perhaps it's because two of the amino acids in spider silk – glutamic acid and arginine – are naturally charged. They are each present in relatively low levels (around 7–9%), but maybe that's enough to give the silk an electric charge? Or maybe the charge is caused by the mechanical stress of forcing the silk out of the spinneret?

Perfectly sensitive

The sensing hairs on the spiders' legs can detect 'sounds that barely register above the random background noise'. What does this actually mean?

It means that they are as sensitive as they can usefully be.

If they were just an ever-so-tiny bit more sensitive, they would continually respond to all the random background noise (whether acoustic or electrical), which would overwhelm the spiders' brains with useless information.

At their current level of sensitivity, they will fire off when, for example, a cloud passes overhead and changes the local electrical field. (For contrast, we humans can't pick up any electrical fields at all with our senses.)

They start by dropping a special anchor silk (from the spinnerets at their back end) onto the ground, to hold themselves in place. They then lift one or both of their front legs up high, for about 5–8 seconds, to sense the conditions (wind, temperature, local electrical field, etc) with specialised tiny sensory hairs.

If the wind is cold, turbulent and speedy, they'll abort the mission.

But if the wind is warm, dry, gentle and under 10 kph, they will get ready, lifting their abdomen (and their spinnerets) up high, almost like standing on their tippy toes. Then, if the local conditions suit the spiders, they will make and immediately release up to 60 silk threads. In one study by Dr Moonsung Cho, these fibres were found to be very thin (0.121–0.323 micrometres, or millionths of a metre) and quite long (2–4 m long). So this gives them a huge amount of surface area relative to their mass – perfect for interacting with any local wind or electrical field.

Then the spiders will break their single anchor line to let go of the ground – and fly off into the heavens!

This is called the 'tiptoe' take-off.

The other take-off method is called 'rafting'. The spiders just hang from a strand in their web, release the ballooning strands or lines when the conditions are right, and again, fly away.

Once in flight, they probably use their legs to help control their path.

SENSING ENVIRONMENTS

The ballooning spiders sense their environment with special sensory hairs, located on their feet. These hairs are called trichobothria.

These trichobothria are exquisitely sensitive to picking up tiny mechanical movements of the air, and sounds that barely register above the random background noise. And as a plus, it seems they can also detect electric fields.

Two scientists, Erica L. Morley and Daniel Robert from the University of Bristol, have studied these ballooning spiders.

In one experiment they placed some of these ballooning spiders on a little insulated stand in the centre of an insulated plastic box, which was also shielded from the Earth's natural electric field. They then varied the strength of the electric field inside the box, so that it was similar to our planet's natural electric field.

And even though there was no air movement inside the plastic box, spiders would lift up their abdomen so that their spinnerets

were elevated and pointing upwards, and start extruding threads of spiderweb.

Amazingly, some of the spiders would actually take off – even though there was no air moving inside the box. And as the scientists increased and decreased (from outside the box) the electric field inside the box, the spiders would rise and fall. The scientists could control the altitude of the spiders, by turning a knob!

Finally, nearly two centuries after Charles Darwin saw these ballooning spiders landing and taking off, our scientists have partly worked out how the spiders make it happen. (Air currents and the Earth's natural electric field, and maybe more.)

Big it up for Charlie D. – he was on the right track, and not just with evolution!

And big respect for all those tiny spiders, recognising the Earth's potential to give them a free ride.

Militarised flying electric spiders!

Of course, for the warmongers, the next obvious step is to militarise this ready army of innocent spiders.

They could 'help' with one 'feature' of our modern fully wired and integrated world – the perpetual, omniscient surveillance that we are all subject to. Our smartphones listen to us and broadcast our location, while millions of tiny cameras watch us everywhere we go.

In 2013, Peter W. Gorham, a physicist from the University of Hawaii, very nicely worked out the physics of how the ballooning spiders could be lifted by the Earth's natural electrostatic field. And he speculated, 'Spiders weighing 100 mg can balloon. That's more than enough to fly a tiny microprocessor and camera.'

Imagine a tiny bio-spider drone, almost too small to see, that didn't need any external power to fly, and could always be spying on you.

I'm an arachnophobe from way back, so that's the stuff of all my nightmares combined!

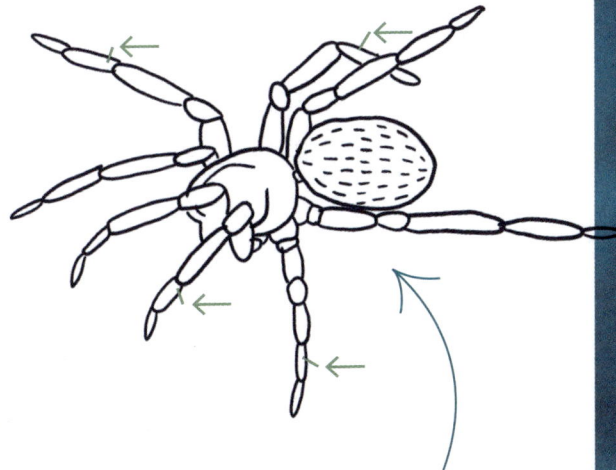

Electron micrograph of adult male Erigone metatarsi and trichobothria, with a close-up view of trichobothrium (inset). Arrows point to the base of trichobothrium. (MT = metatarsus; T = tarsus.)

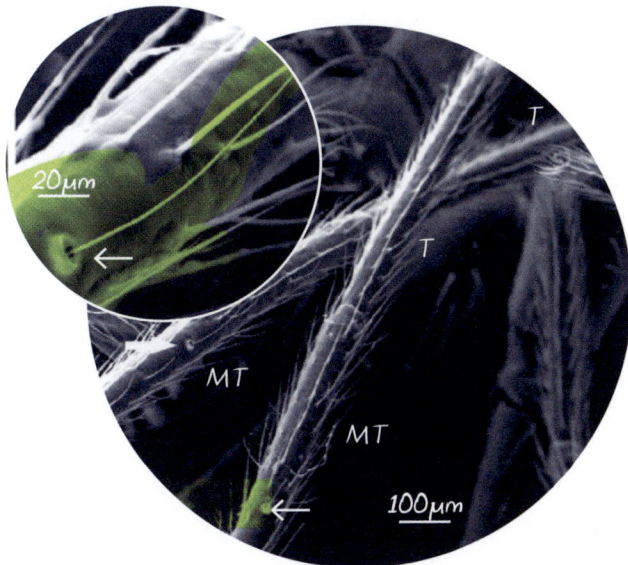

Arrows point to the locations of the tiny hairs (microbothria) that can sense the environment.

20μm

T

T

MT

MT

100μm

Plague of Athens

Smallpox

PAST PLAGUES & CORONAVIRUS

COVID-19

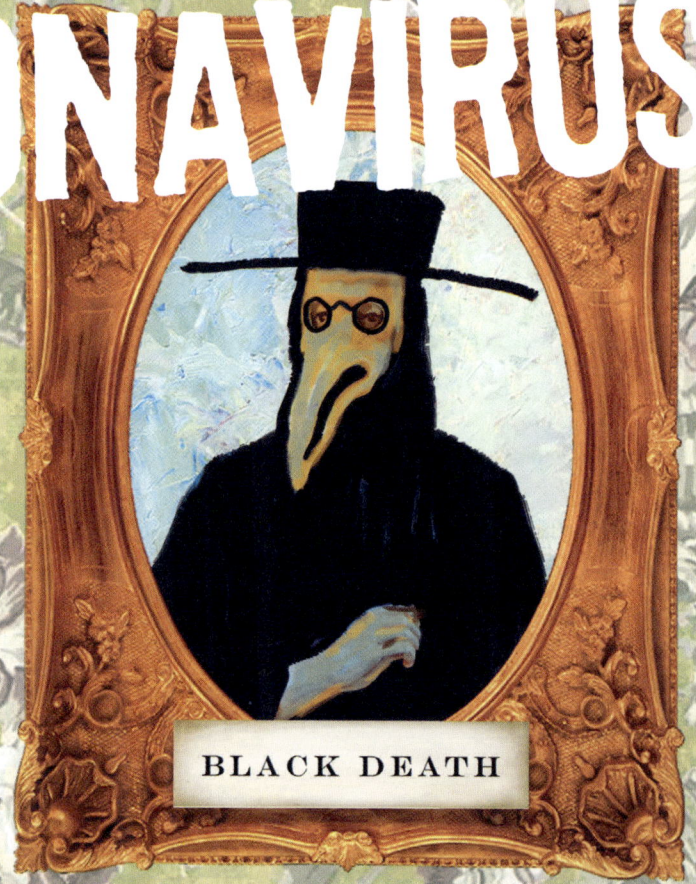
BLACK DEATH

One of Life's Rules – which all decent blues singers know – is that 'old Death never takes a vacation in this land'. Yes, we're all going to die one day – but how?

Well, each year, about 1% of the human population dies. Infectious diseases account for about one fifth of those deaths. But every now and then, a new infection will erupt, seemingly from nowhere, going absolutely 'viral' – and it can kill masses of people. In some cases, the death rate can be 85% of those who get infected.

The word 'plague' is sometimes used to describe such a disease outbreak. One meaning of the word is very specific. 'The plague', or the bubonic plague, refers to the infectious disease caused by the bacterium *Yersinia pestis*, which swept across Europe many times, beginning in the 1300s. Back then, it was called the Black Death, and it killed some 50 million people in the 14th century. It's still around, but infects only about 600 people each year, over the whole world.

Another meaning refers to a massive swarm of insects, as in 'a plague of locusts'.

And finally, 'plague' can mean an outbreak of any infectious disease – whether in humans or animals. The outbreak can be in one location, or worldwide.

EPIDEMIC VS PANDEMIC

There is a normal ebb and flow of mild background infectious diseases coexisting with human populations. And out of that, an epidemic will sometimes arise. An epidemic is an outbreak of an infectious disease, mostly confined to a local area. The word comes from the Greek 'epi', meaning 'upon', and 'demos', meaning 'people'. A recent epidemic was the 2019 Samoa measles epidemic. Out of a population of 200,000, some 5,700 were infected, and 83 died. The cause was massively decreased vaccination rates – for various reasons. While nearby island nations had vaccination rates near 99% (and no cases of measles), in Samoa the vaccination rates had dropped down to about 32% in 2018. Over 90% of the deaths were in children younger than four years old.

Occasionally, an epidemic will break loose to spread across much of the world. Then it's called a 'pandemic', from 'pan', meaning 'all'.

WHY DO EPIDEMICS HAPPEN?

The answer to this question is found in the famous quote from the Mafia movie *The Godfather*: 'It's not personal, it's strictly business.' Sure, we humans might think that we are the crowning achievement of evolution – because we invented literature, nuclear weapons and income tax. But evolution does not have to be perfect, it just has to be 'good enough for the species to reproduce and survive'.

And sometimes, what is 'good enough' for one organism is bad news for another – in this case, humans. So recently, the coronavirus SARS-CoV-2 has reproduced and survived very well indeed, but at a terrible cost to human life.

We humans have shared our ecosystems for hundreds of thousands of years with various types of microbes (bacteria, viruses, prions, parasites, etc). They reproduce and evolve much more quickly than we can. They want to live. And they're everywhere – from the deepest ocean floor to the highest mountain peak. So many viruses exist in our biosphere that if you laid them end-to-end, they would reach 100 million light years – about 50 times more than the distance to the Andromeda galaxy!

The ecosystems we share with microbes include three important elements: the microbes' hosts (the living creatures, or hosts, in which they normally live), vectors (the creatures they use to travel from one host to another, sometimes crossing species), and reservoirs. As an aside, the overwhelming majority of microbes (viruses and bacteria) coexist with us and inside us with absolutely no problems at all. In many cases, we have evolved a mutually beneficial relationship with them. But every now and then, one of these microbes goes rogue. ('Nothing personal, it's strictly business'.)

When talking 'reservoirs', the smallpox virus is a good place to start. Of all humans who died between the years 1000 and 2000 CE, about half were killed by smallpox. In the 20th century alone, it killed 300 million. One reason that the World Health Organization (WHO) managed to eradicate smallpox was that humans were the only reservoir for the virus. (Another reason was that the symptoms were painfully obvious – which meant that it was very easy to identify an infected person.) WHO wiped out smallpox via a global public-health campaign that ran from 1967 to 1977. An Australian, Professor Frank Fenner, was the chair of the Global Commission for Certification of Smallpox Eradication.

But a virus with multiple reservoirs is hard to get rid of. Unfortunately for humanity, the virus that causes influenza has

World's biggest quarantine

In 2020, to try to contain COVID-19, the Chinese central government carried out the largest known quarantine in human history. They placed over a dozen cities under military lockdown, and very effectively isolated *75 million people* from the rest of the world.

many reservoirs – humans, birds, pigs, etc – so we almost certainly will not be able to wipe out influenza. The same seems to be true for SARS-CoV-2, the virus that caused our most recent pandemic of COVID-19.

PLAGUED BY HISTORY

The Plague of Athens (430–426 BCE) helped bring about the end of the Golden Age of Greece. The Athenian historian Thucydides described this disease in superb detail – its signs, symptoms, varying clinical courses, complications, infection rates and death rates. He could also see the big picture, looking at this plague from an epidemiological (a fancy word meaning 'medical statistics') and societal point of view. He wrote about how this plague seemed to have multiple independent origins, how most people were susceptible to it, and how it was linked to overcrowding, war and the breaking down of public-health measures. One in three Athenians died during the plague.

Possibly the first known pandemic was the Plague of Justinian (541 CE). It probably started in Egypt or Ethiopia, raging across the landscape for over a century in several waves and killing about 100 million people, including one quarter of the eastern Mediterranean region's population.

In the mid-1300s, the pandemic of the Black Death rolled around the world. It killed more than 50 million people – and caused fear, panic and massive changes in society. The huge numbers of deaths meant that workers (the peasants) were suddenly hard to come by, leading to social and political tensions. By the end of the 1300s, there had been several peasant uprisings in France, Italy, Belgium and England, and these triggered an end to feudalism in much of Europe.

DR KARL'S Q+A

Where does the word 'quarantine' come from?

It comes from the Italian 'quaranta giorni', meaning 40 days. In Biblical terms, '40', besides meaning that number between 39 and 41, also meant an 'undefined, but really big, number'. So in the Bible, Noah floats on his Ark for 40 days, while Christ goes into the desert to fast for 'forty days and forty nights'.

SMALLPOX EPIDEMIC IN SYDNEY

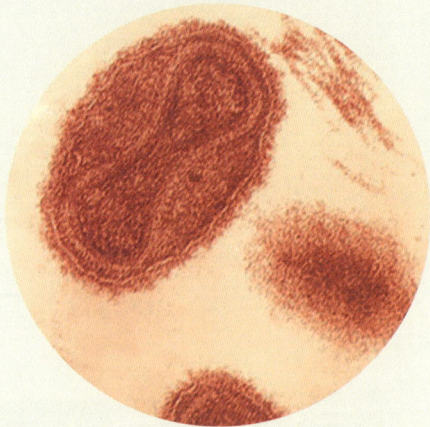

When two separate peoples meet for the first time, so do their germs. It has happened many times across the planet, usually with devastating results, including in Sydney. The British First Fleet landed at Port Jackson in Sydney in January 1788. Smallpox suddenly appeared there 15 months later – but only in the Indigenous population.

It was devastating for the local community. According to historian Alan Moorehead, 'in April 1789 black bodies were suddenly seen to be floating in the harbour and washed up in the coves. Smallpox had struck … by May the disease had spread through all the harbour tribes.' Newton Fowell, a sailor with the First Fleet, wrote that Indigenous people were found 'laying dead on the beaches and in the caverns of rocks'.

On average, since we have kept written records, about 35% of those who caught smallpox died – if there was no vaccination. But protective measures were developed in some parts of the world.

Chinese medical texts from the late 800s CE describe an early form of vaccination, which was called 'variolation'. Smallpox caused visibly raised small scabs (the 'small pox') to appear on the flesh of those infected. The Chinese doctors of the day would scrape off the top layers of the scabs and dry them. Variolation was the blowing of these dried particles (containing the smallpox virus) up the nostrils. It deliberately exposed you to a very low dose of smallpox virus particles to train your immune system to recognise and fight the virus. With variolation, the death rate dropped to only 1–2% (still high, but much better than 35%). The survivors had a far milder form of the disease than if they were infected naturally, and were then immune for life.

Variolation was introduced in Europe by Lady Mary Wortley Montague, the wife of the British Ambassador to the Ottoman Empire, in the early 1720s. She herself had previously been infected with smallpox, and was left with severe facial scarring. Her brother had been killed by smallpox. She came across variolation while in Istanbul, and had it performed on her five-year-old son in 1718. In 1721, after her return to England, she had her four-year-old daughter variolated in front of physicians of the British Royal Court. Later that year, six prisoners in Newgate Prison were variolated with smallpox, and survived, gaining their freedom. As a result, variolation was soon promoted by the British Royal Family.

Seven decades later, most of the Europeans in the First Fleet had already been exposed naturally to smallpox, or had been variolated – either way, they were immune. So the European adults were safe. But what about the unvariolated children who would be born in Australia? What would happen to the unvariolated adults and children among the early white settlers if subsequent ships, carrying smallpox, arrived in Sydney? With this in mind, the First Fleet surgeon John White carried with him a sealed glass bottle of 'variolas matter' – pus from a smallpox victim who had survived – specifically to variolate children born in Sydney. This sounds very reasonable and well-intentioned.

Smallpox did reach Sydney in 1789 and tragically decimated the local native population, while having almost no effect on the European colonists. But how did it get to Australia and then spread so lethally through the Indigenous community? Some historians say it may have been a deliberate act of bioterrorism, introduced as a way to economically kill off a resistant Indigenous population.

We do know that, about 25 years earlier, British forces in Pennsylvania actually did deliberately infect North American Indigenous peoples. The British commander Field Marshal Jeffery Amherst wrote to the British Colonel Henry Bouquet, 'Could it not be contrived to send the small pox among the disaffected tribes of Indians?' Bouquet replied, 'I will try to inoculate the Indians by means of Blankets that may fall in their hands, taking care however not to get the disease myself.' Amherst also wrote to Captain Simeon Ecuyer, 'encouraging him to send smallpox-infected blankets and handkerchiefs to the North Americans surrounding Fort Pitt'. Ecuyer replied: 'I hope it will have the required effect.'

These documents prove that the British in 1763 in North America intended to deliberately infect the local Indigenous peoples with smallpox. Very soon after these letters were written, the Indigenous peoples in that area were stricken by an outbreak of smallpox.

We have zero proof that Australian Indigenous people were deliberately infected with smallpox. But we do know that around Sydney, according to Professor David Isaacs, Clinical Professor in Paediatric Infectious Diseases at the University of Sydney, 'the outbreak killed ... 90% or more of the Aboriginal population'. What a shocking tragedy.

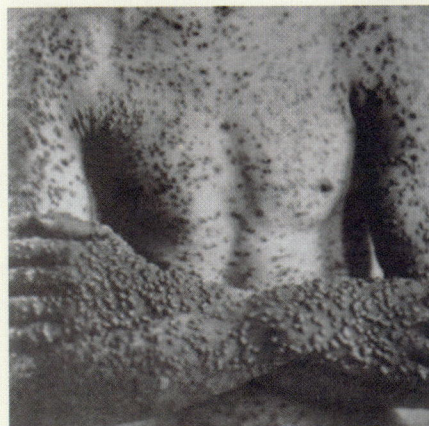

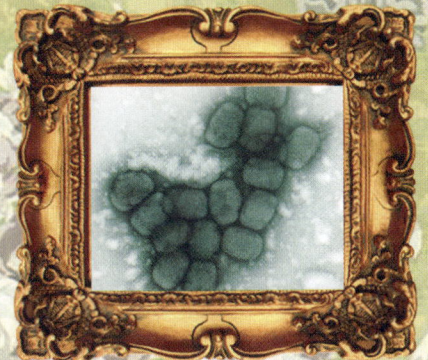

QUARANTINE

You may have heard the story of plague-ridden corpses being tossed over the walls of a besieged city. That city was almost certainly Caffa, now called Feodosiya, in the Crimean region of Ukraine/Russia. Caffa was perfectly located for trade – it had good marine and river connections to Moscow and the Mediterranean, and good land connections (on the Silk Road) between the Far East and Europe. Many cultures met, and traded, there.

But trade wasn't always straightforward. The Mongols and the Genoese (from Italy) had long had an uneasy trading relationship in this city. By 1347, it had degenerated into outright war.

The Mongols attacking the Genoese-held city of Caffa were killed in their thousands when the Black Death infected them. Gabriele de' Mussi wrote: 'The dying Tartars, stunned and stupefied by the immensity of the disaster brought about by the disease, and realising that they had no hope of escape, lost interest in the siege. But they ordered corpses to be placed in catapults and lobbed into the city in the hope that the intolerable stench would kill everyone inside.'

The Genoese merchants fled back to Genoa, carrying the Black Death with them. But they weren't immediately allowed back in. They had to float in quarantine in the harbour. Quarantine was already a well-established concept in the 1300s.

MORE RECENT PLAGUES

Germs, like all life forms, obey the evolutionary imperative to breed, so they will travel anywhere they can survive and multiply. Another great plague, the so-called French Pox of 1494 – which was syphilis – was almost certainly brought back to Europe from the New World (North, Central and South America) by Italian explorer Christopher Columbus.

In the New World, the population around the year 1500 was about 55 million. Over the next century, about 50 million of the New World's inhabitants died – overwhelmingly from diseases imported from the Old World (Europe, Asia and Africa). These diseases included smallpox, bubonic plague, influenza, measles, tuberculosis, diphtheria, scarlet fever, pertussis and a whole range of sexually transmitted diseases.

The American Plague of 1793–98 was caused by yellow fever, which was transmitted by infected mosquitoes. It arose in Philadelphia, where it killed 10% of the population. It then swept

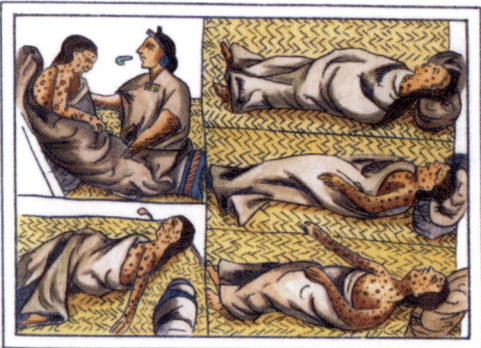

across America during warm weather for the next six years.

Beginning in 1830, a cholera epidemic travelled slowly along the waterways from Asia to Europe. It reached France in 1832. Among those who were infected, the death rate was a terrifying 85%.

The influenza pandemic of 1918–19 swept around the world, just after World War I, and killed 50–100 million people. Interestingly, in some locations, the later waves of infection were more devastating than the first wave.

The HIV/AIDS pandemic has tragically killed over 30 million people since 1981. It was halted by a combination of public awareness campaigns and successful drug therapy.

WHAT CAUSES EPIDEMICS & PANDEMICS?

Dr Anthony Stephen Fauci has been director of the National Institute of Allergy and Infectious Diseases in the USA since 1984, and was a lead member of the 2020 White House Coronavirus Task Force. In a paper he co-wrote, looking at pandemics and epidemics through history, he and his colleagues found over a dozen different factors that were involved, to varying degrees, in these emerging and re-emerging diseases.

International trade, human demographics and behaviour, and an individual's susceptibility to infection were each important. But of course, so were poverty and social inequality, war and famine, breakdown of public-health measures, and technology and industry. For example, thanks to the incredible mobility of air travel, we can get from any capital city to anywhere else on the face of the planet in a day and a half – bringing all our germs with us. Other factors include changing ecosystems, climate and weather; intent to harm (e.g. hurling disease-ridden corpses over walls, or deliberately handing out smallpox-infected blankets to Indigenous peoples); lack of political leadership; microbial adaptation and change; and finally, economic development and land use.

One thing is certain: more epidemics and pandemics will occur in the future. In the same way that every airport in Australia has a firefighting department that is hardly ever used, we should set up a permanent Federal Epidemic and Pandemic Unit that is always ready to respond but is, hopefully, hardly ever used.

Pandemics prune people

Repeated infections sweeping across a population can, over centuries, change that population's genetic make-up.

Perhaps this is why about 10% of all Europeans are highly resistant to HIV (human immunodeficiency virus). Sure, this 10% can be infected with HIV, and the virus can circulate in their various bodily fluids. But the virus cannot enter their cells.

This is because the AIDS virus enters human cells via a receptor called CCR5. This 10% of resistant Europeans have a mutated non-functioning receptor (so they can be carriers, but without symptoms). By a coincidence, the Black Death uses this exact same CCR5 receptor to get into human cells.

As the Black Death swept repeatedly across Europe, anybody with a non-working version of this receptor was more likely to survive and have babies. Perhaps these people expanded in numbers to make up 10% of the European population. Anyhow, that's the 'theory' …

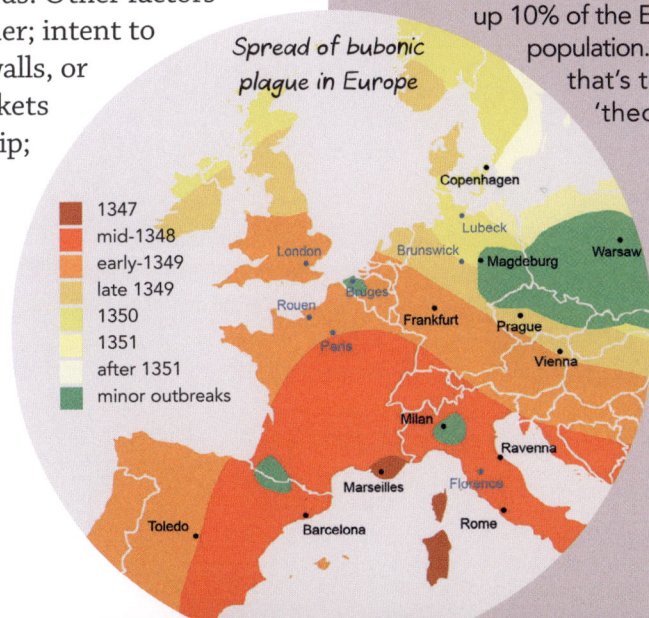

Spread of bubonic plague in Europe

1347
mid-1348
early-1349
late 1349
1350
1351
after 1351
minor outbreaks

Copenhagen, Lubeck, Warsaw, London, Brunswick, Magdeburg, Rouen, Bruges, Frankfurt, Prague, Paris, Vienna, Milan, Ravenna, Marseilles, Florence, Toledo, Barcelona, Rome

RED SKY AT NIGHT

Advice that gets passed down from generation to generation can vary a lot – ranging from very deep and still very relevant, all the way to totally incorrect and dangerous. So you need to be alert.

But here's one saying that, even today, is surprisingly useful: 'Red sky at night, sailors' delight. Red sky in the morning, sailors' warning.' The science behind this folk wisdom is rather neat.

First, the Sun sets in the west. Second, if you can see a red sky after the Sun has set, it tells you there are no clouds for a few hundred kilometres to the west of you. Second again, I'll say the same thing in a different way – because there are no clouds to your west, the rays of the setting Sun can travel all the way to the sky immediately above you. Third, Rayleigh's Law (see page 46) tells us that the longer the distance sunlight travels, the more 'red' that light becomes. And fourth, since tomorrow's weather often comes from the west, 'red sky now' means 'no clouds now', which can often mean 'no rain tomorrow'.

A more poetic way to say it is that when the Sun projects a red sky above us, it is letting us know that there are no clouds between us and the setting Sun (which is a few hundred kilometres to the west, around the curve of the Earth).

Sunlight blocked by clouds

Bland sunset

NORMAL sky at night

SUN (already set in the west)

Air

Air

West

East

Clear air moving towards you

Red clouds

Cloudy air moving away

RED sky at night

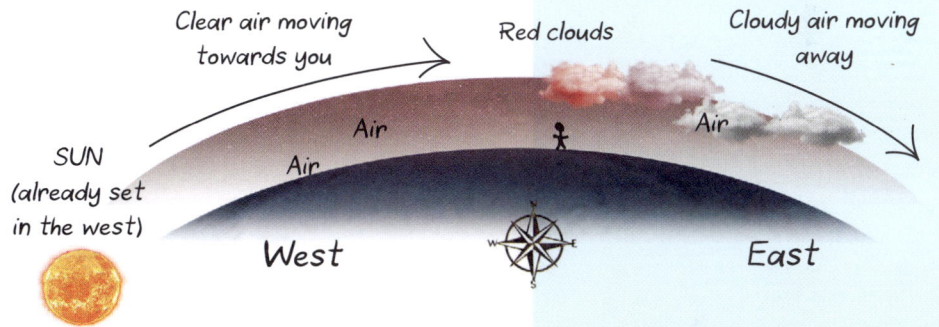

SUN (already set in the west)

Air

Air

Air

West

East

RED SKY 101

This saying about the red sky is pretty old. There's even a version of it in the Bible (Matthew 16:2–3). Jesus Christ himself advises, 'When it is evening, ye say, fair weather: for the heaven is red. And in the morning, foul weather today for the heaven is red and lowering.'

And although this adage is often true, it does come with a few caveats.

Let's look first at just the section about 'fair weather', and totally ignore for the moment the second bit about 'foul weather'.

It works best if you live in the mid-latitudes of our planet, 30–60° either north

Large-scale air circulation

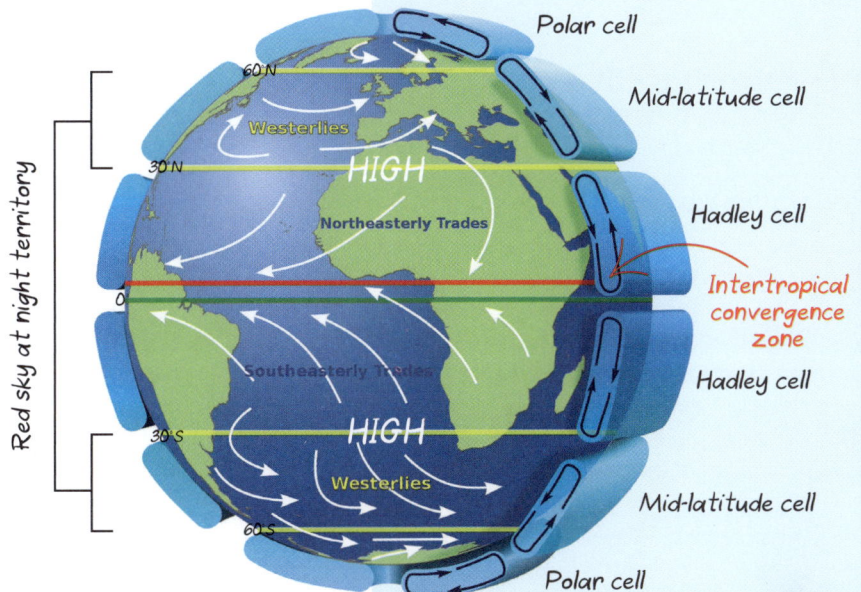

Polar cell

Mid-latitude cell

Westerlies

60 N

30 N

HIGH

Northeasterly Trades

Hadley cell

Red sky at night territory

0

Intertropical convergence zone

Southeasterly Trades

Hadley cell

30 S

HIGH

Westerlies

Mid-latitude cell

60 S

Polar cell

Our atmosphere – just right!

We are very lucky with our atmosphere.

It is thin enough to be transparent. Light, from the stars at night and the Sun in daytime, can shine through it for us to enjoy. (For comparison, if you were on the surface of Venus, which has about 100 times more atmosphere than we do, you would never see the Sun.) We Earthlings get a lover's diamond-scattered sky to appreciate at night, and photosynthesis to grow our plants by day.

On the other hand, our atmosphere is still thick enough to lift water vapour upwards and give us clouds – and rain. (It's a little like how ocean water lifts our bodies and we float, whereas the air around us can't hold us aloft in the same way – unless you're a helium balloon.) If it were as thin as the atmosphere of Mars, there would be hardly any clouds at all.

Earth's atmosphere seems to be just right – within what scientists call the Goldilocks Zone – for life to inhabit it.

or south of the Equator, as Europeans, Americans, Canadians and most Chinese do, and also those in the bottom of South America and Africa, and the bottom half of Australia. In these mid-latitudes, the weather systems usually travel from the west to the east, making the Sun's red projection (after it sets in the west) relevant to tomorrow's weather. However, this proverb does not hold true in the tropics right near the Equator – where the weather patterns don't usually come to you from the west.

We now need three pieces of background information to explain why a red sky tonight predicts fair weather tomorrow.

1. RAINBOW

The first thing to know is that white light is made up of all the colours of the rainbow, running from violet to red. In the late 1600s Isaac Newton darkened a room, let in a skinny beam of light, ran the light beam through a glass prism and – bingo – made his own personal indoor rainbow. (This also gave us Pink Floyd's famous album cover for *The Dark Side of the Moon*.)

Newton proved that 'hidden' inside the white light that is all around us is this lovely potential rainbow of all the possible colours. (This is part of the reason that the LGBTQ community chose the rainbow flag as the perfect symbol for its cause – it represents vast diversity as well as inclusivity in the one motif.)

And right at one end of the spectrum, we find the colour red. If we have a big picnic planned for tomorrow, this is the colour we want to see in the evening sky tonight.

2. RAYLEIGH'S LAW

But how does Nature give us the red colour in the sky, so we can predict tomorrow's fine weather? This brings us to the second background thing we need to know: Lord Rayleigh's Law of Molecular Scattering. Relax – it sounds complicated, but we'll take it slowly.

'Scattering' has a special meaning for physicists. It doesn't just mean 'tossed in all directions'. Instead, it means that something (e.g. light) changes direction because of some kind of 'interaction'. In another example, an electron might scatter off a proton, and change its direction of travel.

FACT CHECKING FOR OBSESSIVE PEOPLE LIKE ME

I can't go any further, until I get these whinges off my chest.

1) Sunlight is white, not yellow. And the colour of the Sun is white, not yellow. Why do primary school kids colour in the Sun as yellow? Because they have white paper, and a **white** Sun does not stand out on **white** paper. If we all used black paper, then we could have painted the Sun its correct colour (which is white!) when we were in primary school, and not grown up wrongly believing that the Sun is yellow.

2) I love Pink Floyd, but there is no 'Dark Side of the Moon'.

That's like talking about the Dark Side of the Force. There is no 'side' of the Moon that is always in 'darkness'. At almost all times, one half of the Moon is in sunlight, while the other half is in shadow. The half of the Moon that is in shadow/darkness keeps on changing, as the Moon goes through its very slow rotation that takes about 28 days. (By the way, the only time that both sides of the Moon are in shadow is during a total eclipse of the Moon – when the Moon goes into the Earth's shadow.)

Dark Side of the Moon? That's 'emotion', not 'optics'.

However, because the time taken for the Moon to orbit about the Earth – 28 days – is the same as the time taken for the Moon to spin or revolve once on its own axis – 28 days – one side of the Moon is always visible to us here on Earth, while the other side is not visible to us Earth-bound non-astronauts. That's the Unseen Side of the Moon – which is sometimes in darkness, and sometimes in light. But the Unseen Side of the Moon is in darkness half the time, and in sunlight the other half of the time.

In our atmosphere, when sunlight interacts with a molecule of oxygen or nitrogen, it is 'absorbed', and a little later it is 'emitted' with two changes.

The first change is that it's mostly emitted at right angles to the direction that it came from. (That's a little irrelevant to this story, but it does explain why the sky is blue.)

The second change is that once it has been emitted, the colour is just a little more blueish – compared to its colour before it was absorbed. (Not strictly correct, according to the Clever People in Physics, but close enough to the truth for the likes of you and me.)

Rayleigh's Law also says that 3.2 times more of this blue light is likely to be diverted away from its original path than red light.

3. AIR MASS

The third thing we need to know is how much air you are looking through when you look into the distance. We call this the 'Air Mass', or AM.

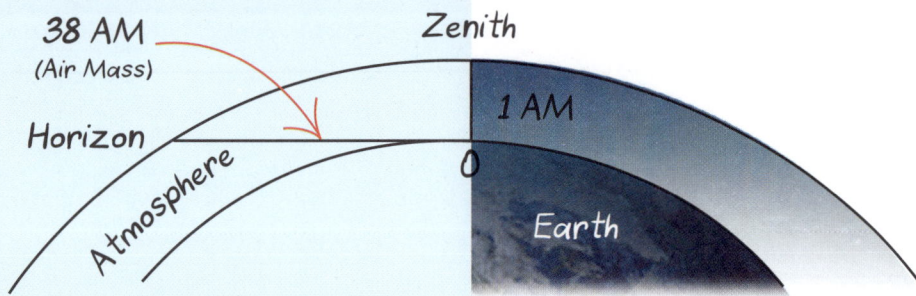

38 AM
(Air Mass)

Zenith

Horizon

Atmosphere

1 AM

0

Earth

If you look straight up to the heavens, the Air Mass between you and outer space is defined to be '1 AM'.

But when you swing your gaze downwards by a right angle (90°), and look at the Sun as it is setting on the horizon, you are looking through a whole lot more air molecules. How many more? Actually, about 38 times more.

This is because when you look at the horizon, you are looking through many more kilometres of air, and, furthermore, this air is denser than the air at higher altitudes. (By the way, these molecules in the air are about 78% nitrogen, 21% oxygen, 1% argon – and bits and pieces of other gases, including carbon dioxide at 0.04%.)

WHY THE SETTING SUN IS RED

Combining the above three factors explains why the setting Sun can turn red as it just kisses the horizon.

When you look at the Sun on the horizon, you're seeing it through 38 AM (or 38 times the Air Mass, or 38 times more molecules) than when the Sun is directly above you. This means that the direct light from the setting Sun is scattered 38 times more than when the Sun is directly above you. But on top of this, Rayleigh's Law tells us that while all the colours get scattered, blue light gets scattered 3.2 times more.

So what about the remaining light that lands inside your eyeball?

This remaining light is overwhelmingly red – the red is what is left over, after the other colours get diverted by the scattering.

Okay, so that's why the setting Sun is red in colour.

RED SUN – RED SKY

But why is the saying about a red sky, not a red Sun? And what about the fine weather supposedly coming to visit you the day after you see this phenomenon?

Well, when the Sun drops below the horizon, the sky does not immediately go black. The Sun is now around the curve of the Earth, and its light has to travel even further to get to your part of the world. This is the lovely time known as twilight. The Sun hasn't

stopped descending – in fact, it's still dropping lower and lower below the horizon.

The reason the sky hasn't gone black is that the light from the Sun is being gently bent by the atmosphere. As the sunlight travels on its journey from below the horizon (around the curve of the Earth), it travels through even more air molecules – even more than 38 Air Masses. So there's even more scattering, and even more of the blue colour is bent away out of the beam of sunlight. That blue light has gone somewhere else. All that is left is red light – which will illuminate the sky above you.

But you get to see the red sky – and this is the important part – only if there are no clouds between the Sun and you. Otherwise, the clouds will block the sunlight, and you won't get that lovely red twilight sky.

And what's the direction of the setting Sun? Yes – the west.

And in mid-latitude places, where do the weather systems like stormy rain clouds come from? Yes, same again, the west.

So if you've got a very, very red sunset, it means there are no clouds to the west for a few hundred kilometres. Because weather systems generally don't move very quickly, if there are no clouds to your west at sunset, that cloudless sky will be directly above you sometime tomorrow.

In general, no clouds means no rain, and fine weather.

BUT HOUSTON, WE STILL HAVE A PROBLEM ...

So now, it all makes sense – 'red sky at night, sailors' delight'. The Sun that has already set is telling you that there are no clouds to the west by making the westerly sky red. No clouds tonight probably means no rain tomorrow.

But first, this does not work if the local weather pattern happens to be coming in from the south, the north or the east.

Second, some parts of the planet (e.g. the tropics) have weather systems that generally do not come from the west, as we saw in the Air Circulation diagram on page 45.

It's taken me quite a few years just to understand the first half of this saying. Forget the other half – 'red sky in the morning, sailors' warning' – I can't explain that yet. But one day (and I hope that day comes soon), when I'm fully convinced, I'll make sure to let you know.

GREEN ROOM

Can you look directly at the setting Sun?

When you look at the setting Sun, you're looking through 38 times more air than when it's overhead. So, most of the light from the setting Sun heading in your direction gets bent away from you – it does *not* get into your eyeball. This means you can bear to look at the setting Sun with your naked eye – and not damage your eye.

Of course, do this only in very brief bursts, and very cautiously.

BLACK HOLES –
CLOSE & MISSING

Around mid-2020, astronomers announced the discovery of HR 6819, a triple star system that includes the nearest black hole to Earth — only 1,120 light years away. This discovery brings the number of black holes discovered in our galaxy, the Milky Way, to a few dozen. That was the good news.

The bad news is that black holes are given boring and robotic names, even though they are amazing.

And the really bad news is that we still haven't found the other 100 million to 1 billion black holes that should be out there.

How come?

BLACK HOLE HISTORY

Let's start at the beginning. How does a black hole form?

A black hole can form in a few different ways, but they always involve 'gravitational collapse'. This is where the gravity of an object makes it collapse into itself.

One pathway is that a supermassive cloud of gas and dust coalesces (or shrinks) under its own gravity to form a supermassive black hole. It doesn't even go through the stage of igniting into a star. We think this is how some of the biggest black holes formed, way back in the early days of the Universe. These weigh in at billions of times the mass of our Sun (which we call one 'Solar Mass'). Alternatively, the cloud of gas and dust could turn into a black hole of 100,000 Solar Masses, and then continue to grow by eating anything that came within reach.

Another pathway by which a black hole forms is when a very massive star collapses after running out of nuclear fuel to burn. So a star of 20 Solar Masses could lose 15 Solar Masses in its final stages of life, before ending up as a black hole of 5 Solar Masses.

We still haven't worked out all the details. For example, did black holes with millions of Solar Masses get there by a single collapse into something close to their final mass? Or did they catastrophically shrink into something half their final mass, and

then get more massive by swallowing any matter that got close enough? We don't know – yet.

Regardless of how it gets there, a black hole has its matter 'stored' inside a volume of zero, i.e. no size.

Often, a black hole is surrounded by a spiralling disc of collected stuff – star dust, tiny rocks, asteroids, alien spaceships (who knows?) and more. It's called an 'accretion disc', and it's outside the 'event horizon' (more on these on pages 86–89).

In 1784, an English clergyman, John Michell, put forward the concept of a star so massive that even light could not escape. (He is now considered 'one of the greatest unsung scientists of all time', thanks to his insights into fields as widely varying as optics, geology, astronomy and gravitation.) But back then, a popular scientific guess was that light was made of particles (this is only partly correct). In 1915, Albert Einstein developed his General Theory of Relativity, which is really a theory of gravity, which gave physicists the framework for a deeper understanding. Soon after, German astronomer Karl Schwarzschild used Einstein's work to describe the gravitational field of a mass that had no size. Many other scientists added to the theoretical knowledge of black holes – Georges Lemaître, Subrahmanyan Chandrasekhar, David Finkelstein, Roger Penrose and Stephen Hawking.

FINDING BLACK HOLES

It took longer for the astronomers' experimental measurements to catch up with the theory. The reason they eventually found them was that while black holes might be 'black', they are not totally invisible.

The first black hole to be discovered, Cygnus X-1, was found in 1964. It's about 6,000 light years away. It's part of an X-ray binary system – which means that two stars are orbiting each other, and the system is emitting X-rays. Cygnus X-1 has a black hole and a blue supergiant variable star orbiting each other fairly closely (about 20% of the distance between the Earth and the Sun) every 5.6 days. The blue supergiant has a surface temperature around 31,000°C (about six times hotter than the surface of our Sun), and is about 20–40 Solar Masses. The black hole (14.8 Solar Masses) is sucking matter from the blue supergiant onto an accretion

The blue supergiant is on the left. It has been strongly distorted (into a teardrop shape) by the gravity of the invisible black hole on the lower right. The accretion disc surrounds the invisible black hole.

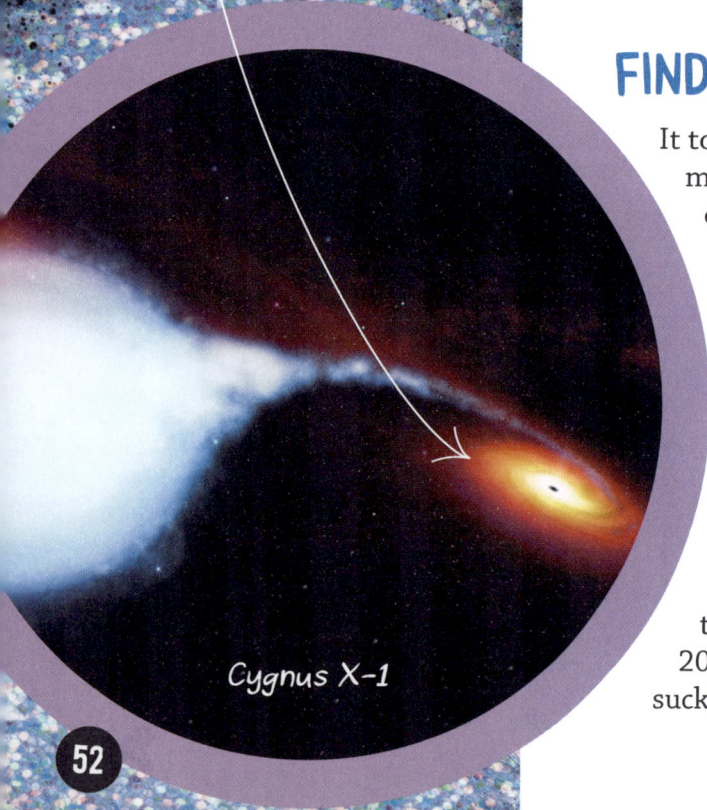

Cygnus X-1

disc orbiting around the black hole's equator. (Yes, I know – how can something that has no size still have an equator? Dunno, but that was the easiest way for me to describe it.) In fact, the gravity of the black hole, even at this distance, is so intense that it has distorted the blue supergiant into a teardrop shape. The infalling star stuff starts getting so hot (millions of degrees), it gives off X-rays. These X-rays are what we measured back in 1964, when we launched a sounding rocket to look for sources of X-rays in the heavens.

As part of the process of the matter falling inwards, it will often end up in a spiral around the equator of the spinning black hole. Most of this matter will definitely fall into the black hole. Some of the matter might get squirted out as an astrophysical jet – also called a cosmic jet, or a relativistic jet if it's going fast enough. (Sometimes, half falls into the black hole, and the other half gets squirted out.) But getting back to the stuff that will journey into the black hole – this process takes time. It's like pulling the plug in a bathtub full of water – the water will definitely empty down the drain, but it doesn't happen straightaway.

This matter that falls in from a companion star leads to the formation of an accretion disc, which spirals in towards the event horizon. The stuff within this accretion disc is moving very quickly and often rams into other stuff within the disc. It's these collisions that give off radiation that we can detect. This is how astronomers have found most of the few dozen black holes in our Milky Way galaxy.

So that is one way to find black holes.

Another way to detect a black hole is if it is conveniently in orbit around a regular shining star. The black hole and the shining star orbit their common centre of gravity. We can't see the black hole because nothing can escape it. But if we very carefully observe the shining star, we can see that sometimes it comes towards us, and sometimes it goes away from us. This then changes the observed frequency of the regular star's light – higher when it's coming towards us, and lower when it's going away. This is the well-known Doppler shift, which is used in police radar to measure the speed of an approaching, or departing, car. So by measuring the shining star's Doppler shift, we can tell that it is oscillating back and forth around an unseen object. The fact that the regular star is orbiting something proves that the 'something' exists, even if we can't see that 'something' directly.

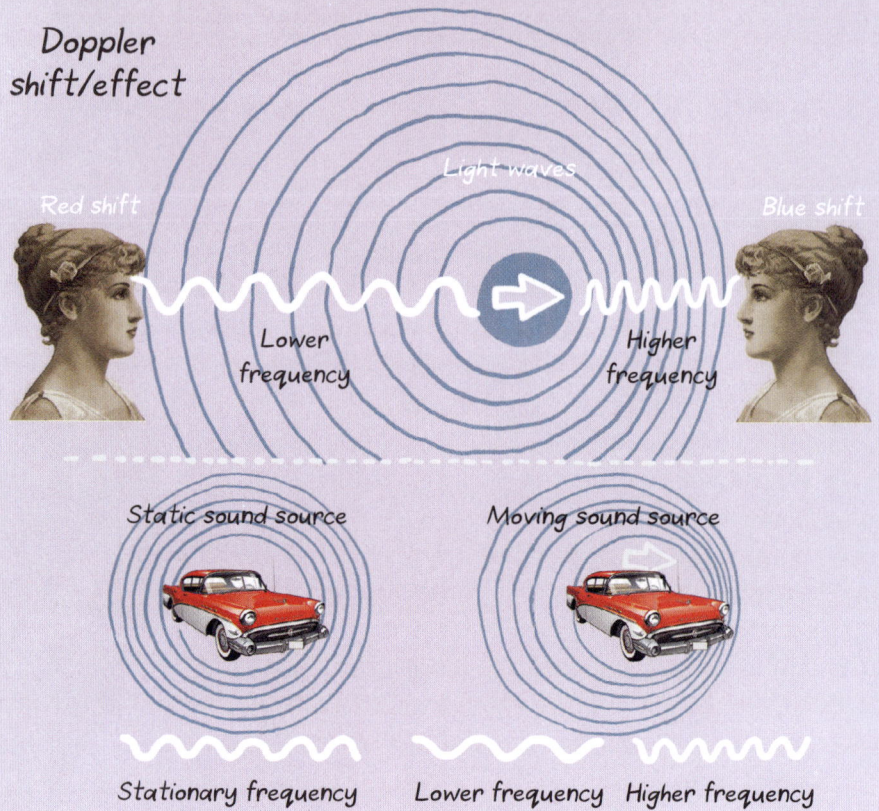
Doppler shift/effect

Light waves

Red shift

Blue shift

Lower frequency

Higher frequency

Static sound source

Moving sound source

Stationary frequency

Lower frequency   Higher frequency

The HR 6819 triple star system. The inner binary pair are the black hole (red orbit) and a blue giant star (green orbit) in a fairly tight orbit around their common centre of gravity. The triple is completed by another star (longer green orbit) in a much bigger orbit around the binary pair – and the common centre of gravity of all three stars.

This Doppler shift was the method used to find the black hole that is (as of June 2020) the closest known black hole.

NEAREST BLACK HOLE

Now, it might be a surprise to you, but when you go out at night and gaze at the multitude of visible stars in the sky, rather than them all being single stars, about 40% are actually two stars in orbit around each other. And some of them are even three stars orbiting one another.

If you go outside on a cloudless and moonless night, away from city lights, and stare into the constellation called Telescopium, you can actually see, with the naked eye, the triple system known as HR 6819. It's quite faint, about as bright as the planet Uranus. For it to be visible, you need to be south of 33°N in latitude (which we are in Australia).

Making up two thirds of the triplet star system of HR 6819 are a black hole and a blue giant star – a pair of celestial objects in a tight orbit around their common centre of gravity. The black hole

weighs in at around 4.2 Solar Masses, while the blue giant is about 6.3 Solar Masses. They take about 40.3 days for a complete orbit. That inner blue giant is expected to expand enormously in a few million years. It will probably undergo interesting energetic events with the black hole, as its surface expands and gets closer to the black hole ...

Completing the triplet is a third star. It orbits the black hole and the blue giant, but in a much wider orbit. This outer star is spinning so quickly that it bulges at the equator.

That black hole is the remnant left over from the death of a massive young star (perhaps 20 Solar Masses), when it exploded as a supernova approximately 15 million years ago. The other two stars are hot (some 15,000°C, about three times hotter than our Sun) and young (about 50–140 million years old).

And yes, that black hole is the closest (currently) known black hole! It was also one of the very first black holes discovered that was not having violent interactions with its environment, such as giving off X-rays. Instead, it was found with the Doppler shift method.

MISSING 1 BILLION BLACK HOLES!

Now, the Milky Way galaxy has been around for well over ten billion years.

Over that time, many massive stars have been born in stellar nurseries, lived a fast and furious life, and then collapsed into a black hole. Sometimes, they completed their whole life cycle in less than 20 million years. Our Sun is different – it has been around for 4.6 billion years, and has another 5 billion years to go before it goes through the red giant stage into the final white dwarf stage.

Yes, big stars live fast and die young, while smaller stars live slowly and die old.

At the moment, there are about 300 billion stars in the Milky Way, which is at least 12 billion years old. So when you do the calculations, you would expect that somewhere between 100 million and 1 billion massive stars in our Milky Way should have turned into black holes.

But so far, we have found only a few dozen of them. Almost certainly, the vast majority of these missing black holes are not gobbling up lots of mass. They're 'resting'. So they're not emitting radiation from the accretion disc swirling around them - which makes them fairly invisible to our current technology. Or maybe

DR KARL'S Q+A

Where did the name 'black hole' come from?

In 1784, John Michell used the phrase 'dark star' for a celestial body so massive that not even light could escape its gravitational field.

In the early 1960s, the physicist Robert H. Dicke supposedly compared such an object to the 'Black Hole of Calcutta', a notorious dungeon from which inmates supposedly never left alive. (Misconception Alert – it's wrong. Actually, the dungeon got its name after an Indian rebellion that occurred on 20 June 1756, where 123 of the 146 British prisoners imprisoned there died overnight. The room was only 4.3 x 5.5 metres, and intended for only two or three prisoners at a time.) In 1963, the phrase 'black hole' was used in astronomy stories in both *Life* and *Science News* magazines. In January 1964, science journalist Ann Ewing wrote an article, '"Black holes" in space', reporting on a meeting of the American Association for the Advancement of Science.

But the phrase became widely used only after December 1967, when a student supposedly suggested it to the physicist John Wheeler at one of his lectures. Wheeler started using it (because of its brevity and 'advertising value'), and it then spread rapidly.

And no wonder – 'black hole' is a great name!

55

Hawking loses bet

Back in 1974, Stephen Hawking (black hole dude, above) and Kip Thorne (gravitational wave dude, below) had a friendly scientific bet on Cygnus X-1. Hawking reckoned it was not a black hole. The stakes were magazine subscriptions.

Hawking conceded in 1990 that the observational data was strong enough to show that Cygnus X-1 was indeed a black hole, and paid for the magazine subscription for Thorne.

they are just a single black hole, and not part of a binary system, which makes them even harder to find.

Surprisingly, although we only just realised that HR 6819 includes a black hole, it was actually observed way back in 2004. The data from that observation was reanalysed in 2020, only because another group had recently found a similar black hole, called LB-1. The LB-1 astronomers analysed the Doppler shift of its companion star to confirm it was a black hole. This rang a bell in the memory of the group's chief scientist, Dr Thomas Rivinius. He thought, 'Wait a second. I have something in my drawer of unused data that looks pretty much like [LB-1].'

Yep, they found another black hole, HR 6819, in his drawer. (At least, it's much better than finding a skeleton in your closet.)

This suggests that once we look more carefully, we will find a lot more black holes. But the black holes that are not either influencing a nearby star or gobbling in matter – well, they will be much more difficult to find. On the other hand, if there are 100 million of them, and if they are evenly scattered throughout our Milky Way galaxy, the nearest should be just 30–40 light years away – making detection much easier.

A 'black cat on a black night' is really hard to find. With our current technologies, that pretty well sums up the challenge of finding black holes.

PLANET NINE

Some astronomers claim that there is an undiscovered planet in our Solar System, way out past Pluto. They reckon that it's very massive – 5–10 times the mass of the Earth – and takes 10,000–20,000 years to do a complete orbit around the Sun. They call it 'Planet Nine'.

Their evidence is that a whole bunch of objects (comets and the like, way out past Pluto, in the Kuiper Belt) seem to have been thrown into elliptical orbits. Their guess is that the so-far-undiscovered Planet Nine disturbed these objects with its gravity – and tossed them towards the Sun. These disturbed objects with very elliptical orbits are called extreme trans-Neptunian objects.

We have been looking for this planet for nearly two decades – but if it exists, it is very far away and so very hard to see. Sure, it might be 40,000 km in diameter, but because it's so distant, it would look extremely tiny to us.

Imagine, instead, that this object is not a planet, but a very low-mass black hole. In that case the event horizon would be only a few centimetres across – and that would make it almost impossible to find with our current technology.

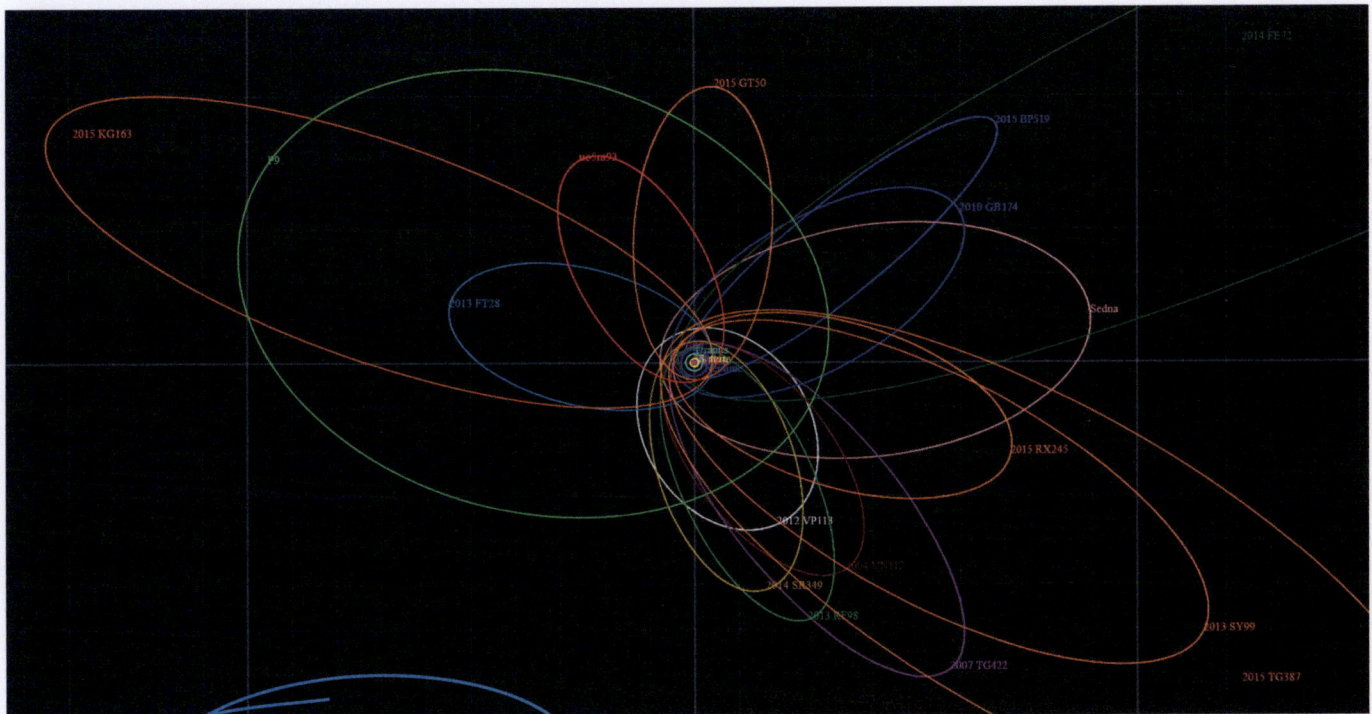

The orbits of 14 known extreme trans-Neptunian objects are the large elliptical orbits. The hypothetical Planet Nine has the green orbit. The set of small concentric circles near the centre represents the eight known planets, from Mercury to Neptune.

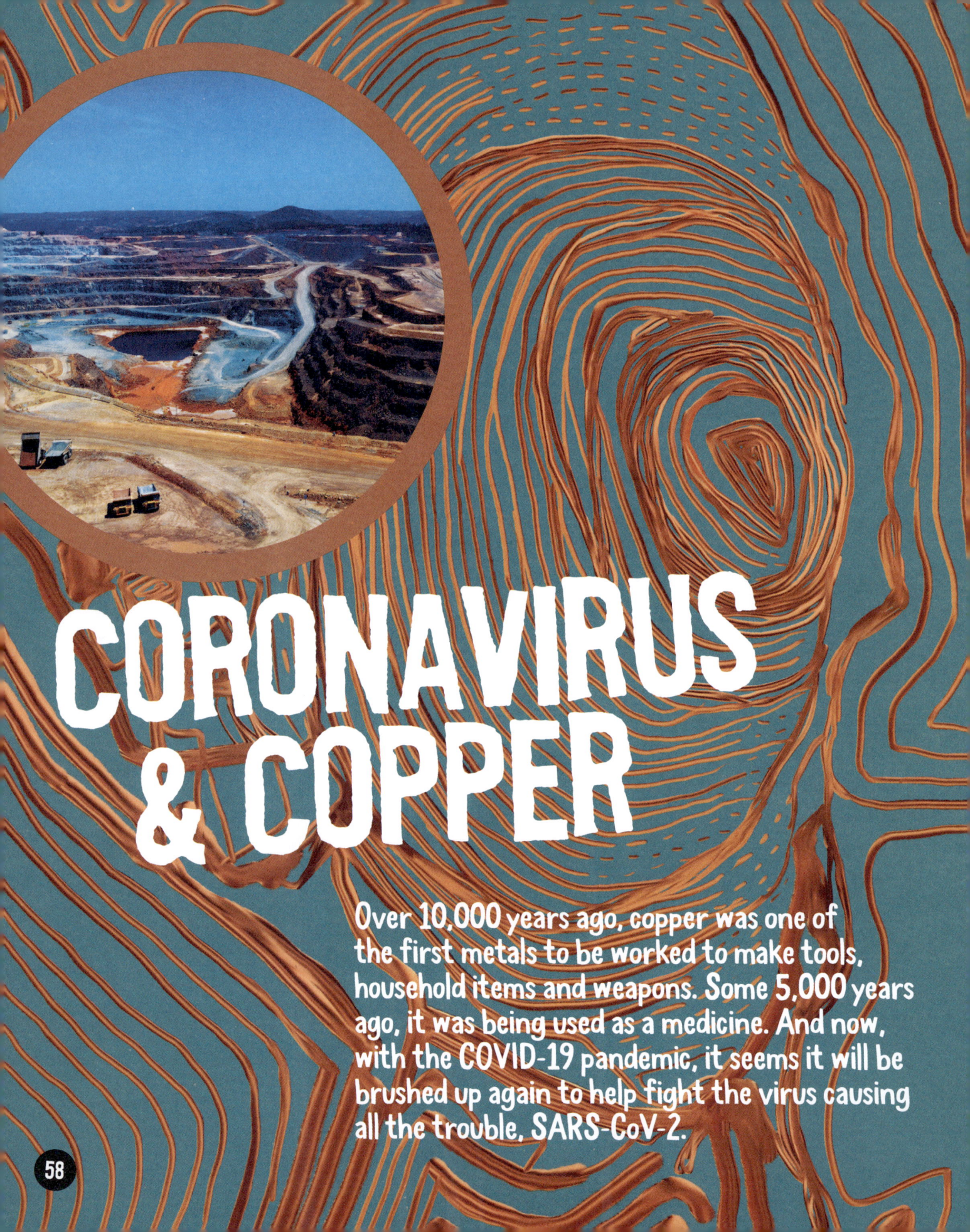

CORONAVIRUS & COPPER

Over 10,000 years ago, copper was one of the first metals to be worked to make tools, household items and weapons. Some 5,000 years ago, it was being used as a medicine. And now, with the COVID-19 pandemic, it seems it will be brushed up again to help fight the virus causing all the trouble, SARS-CoV-2.

COPPER 101

Today, we mine for copper ores in huge open pits. The ores typically contain 0.4–1.0% copper.

But in the old days, you could find lumps of pure copper lying around on the ground. (Gold is another such 'native metal', also found in pure lumps.) The biggest such lump of copper was found in Michigan, and weighed 420 tonnes.

Copper was the first metal to be extracted from ores (around 7,000 years ago), the first metal to be cast in a mould (about 6,000 years ago), and the first metal to be deliberately blended with another metal to make the alloy bronze (about 5,500 years ago).

And 3,800 years ago, copper was the subject of the world's oldest complaint letter. This letter was written in cuneiform on a baked clay tablet. Basically, in so many words, the letter was saying, 'Stop ripping me off!' The writer protested about the low quality of copper ingots being offered, that his agent was being treated rudely, and, worst of all, that he had neither copper ingots nor his money anymore.

Copper gets its name from the island of Cyprus, where the Romans mined huge quantities. So the metal was originally called 'aes cyprium', which meant 'metal of Cyprus', which eventually got shortened to 'cuprum' in Latin.

Even though we have been using copper for at least 10,000 years, the vast majority (95%) of all the copper ever mined has been extracted only since 1900 CE. There's about 100 trillion tonnes of copper in the top kilometre of the Earth's crust. At the current rate of extraction, it should last about five million years. However, copper is beautifully recyclable with no loss of quality – so 80% of all the copper ever mined is still being used today.

Copper is mostly used for electrical wire (60%), plumbing and roofing (20%) and industrial machinery (15%).

Copper can exist in a variety of 'oxidation states' – copper (I), copper (II), copper (III) and copper (IV). This archaic term was coined by the French chemist Antoine Lavoisier in the 1700s. Today it refers to the number of electrons that an atom has lost. The fact that copper can give up varying numbers of electrons means that it has the capacity to be involved in many different chemical reactions. And this varied chemical reactivity might be related to its involvement with 'life'.

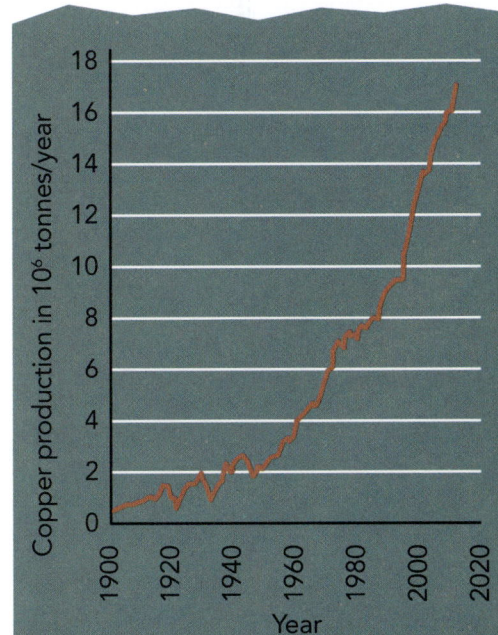

4cm

Copper production in 10^6 tonnes/year

Year

The Egyptian god Horus offers eternal life (the ankh) to Rameses II.

Bronze vs brass

After years of guessing, I have finally worked out how to remember the difference between bronze and brass. Each is an alloy of copper. This memory aid is based on Opposite Day (from *Play School*).

Bronze has the letter 'z', which is the first letter of the symbol for the element zinc, Zn, but it does *not* contain any zinc – it usually mixes copper with tin. (See, Opposite Day!)

Brass has the letter 's', which is the first letter of the symbol for the element tin, Sn, but it does *not* contain any tin – it mixes copper with zinc. (More Opposite Day.)

If only it were the other way round! This 'opposite-day-thinking' helps me, but I don't know if it will help anybody else.

COPPER – LIFE & DEATH

Copper is an essential trace element for plants and animals. We humans carry 1.4–2.1 mg of copper for each kilogram of body weight. Copper is absolutely essential for the body's metabolic processes, such as where it uses glucose to make energy (ATP, see page 95). In this case, copper is found inside the enzyme 'cytochrome c oxidase', located inside mitochondria, a strange 'mini-cell' that lives inside our cells. Copper is also essential for structural and chemical processes, such as the manufacture of collagen and melanin in the human body. Collagen is one of the major structural proteins in our connective tissues, and indeed, is the most abundant protein in mammals, accounting for about 30% of all the proteins in a mammal. Melanin is a general name for a group of natural and essential pigments in most creatures – such as in skin, hair and eye colour, and in various parts of the brain, the adrenal glands and so on.

So copper is absolutely essential to promote some processes in living creatures, but it also has the power to interfere with other biological processes.

Biofouling is the build-up of various microbes, algae, plants and even small animals on surfaces that are wet. Copper has been used for centuries on ships' hulls to prevent mussels and barnacles growing on them – which can increase the drag on a hull by as much as 60%. Copper is also used in aquaculture to stop biofouling.

Copper is mentioned in the Edwin Smith Papyrus, one of the oldest known medical documents. The information in this Egyptian document goes back 5,000 years. In fact, as part of their reverence for copper, the Egyptians specifically assigned the 'ankh' symbol in their hieroglyphs to represent both 'eternal life' *and* copper.

Some 3,500 years ago, Chinese doctors used copper coins to treat diseases of the heart, stomach and bladder.

The Phoenicians, Egyptians and Babylonians each had large military forces. The soldiers and sailors would save the metal shavings after they sharpened their bronze swords. It was common practice to put these shavings (containing copper) into battle wounds, to speed healing and to reduce infections.

Civilisations such as the Romans, Aztecs and Greeks had medical uses for copper – 'headaches, burns, intestinal worms, and ear infections ...'

And there's long been the folk-culture knowledge that if a

family drank water from copper vessels, they were less likely to suffer diarrhoea when a gastro (stomach) bug swept through the area. (Aren't we lucky that we use copper pipes to carry our drinking water?)

In the cholera epidemic of 1832 in France, it seemed that copper workers were relatively immune to the disease. Also in France, the wineries applied Bordeaux mixture (copper sulphate and slaked lime) to vines to stop fungal attack – and still do.

Professor Bill Keevil (of the University of Southampton in the UK) has been studying the effect of copper on bacteria and viruses for a quarter-century. He has shown that mere contact with copper kills bacteria, such as the one that causes Legionnaire's disease, or the Methicillin-Resistant Staphylococcus Aureus (MRSA) superbug. It also inactivates viruses such as the coronavirus that caused Middle Eastern Respiratory Syndrome (MERS) in 2012, and the influenza virus that caused the swine flu (H1N1) pandemic in 2009.

HOW DOES COPPER KILL GERMS?

How copper kills viruses and bacteria is still not fully understood. But this antimicrobial power seems to be related to copper's position in the periodic table. It's in the same column (group 11) as gold and silver, which also have some degree of antibacterial power. However, silver has antimicrobial activity only when it is wet, while copper is active both when it is wet and dry. This ability to smash microbes seems to be related to copper's single free electron in its outer shell of electrons.

Just as a reminder, an atom looks a bit like a grape in the middle of an empty football field. The grape is the heavy nucleus where most of the mass of the atom is concentrated. Around it is a huge amount of empty space, and, finally, out where the crowd sits, are clouds of electrons in a shell, orbiting that central nucleus. It gets complicated, with some of the

GREEN ROOM

Weakly radioactive copper

Copper is the 29th element in the periodic table. By a ridiculous coincidence, it also has 29 isotopes – an astonishingly high number. (An isotope of any element always has the same number of protons in the nucleus, but varying numbers of neutrons. Most elements have only a handful of isotopes.)

Each isotope of copper has 29 protons, but the number of neutrons varies between 23 and 51. Only two of these isotopes are stable. The rest are weakly radioactive (i.e. not dangerous to us), with half-lives between 75 billionths of a second and 61.83 hours.

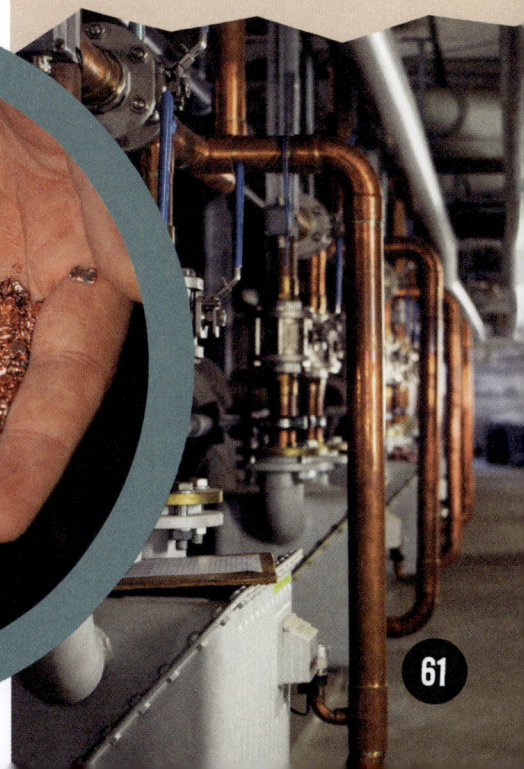

concentric shells having sub-shells that are not filled completely. But regardless of what's happening in the inner electron shells, copper has a single free electron in its outermost shell.

Consider a single bacterium, or virus particle, sitting on the surface of the copper metal. It seems that atoms of copper physically drift, or diffuse, from the surface of the copper into the cell membrane of a bacterium, or into the outer coating of the virus particle. The copper atoms/ions then set off various chemical reactions, some of which manufacture hydrogen peroxide (commonly used as bleach). The hydrogen peroxide blasts holes through the cell membrane of the bacterium, or disrupts the outer coating of the virus. This destroys the germ's ability to infect us.

There are probably other chemical pathways of attack. Copper can interfere with the roles of the metals iron and zinc, oxidise chemical groups in proteins, make hydroxyl radicals, and more.

It's still not fully confirmed, but it also seems copper has another action, deeper inside the bacterium or virus. It appears to damage the genetic material of the bacterium or virus – its DNA or RNA. This might stop the rise of mutations that could make the germs resistant to copper.

Copper's antimicrobial effect is long-lasting. Professor Keevil's team checked out various hand railings inside Grand Central Terminal in New York, which were installed over a century ago. Mechanically, the copper handrails still have structural integrity – they are as good today as they were back then. And even better, that century-old copper still has its antimicrobial effects.

Now, on one hand, I have a long-standing love affair with stainless steel. Stainless steel is shiny with a beautiful mirror finish – and I'm a sucker for sparkly things – while copper tarnishes from a brown to a green finish. But looks aren't everything.

Stainless steel has many micro-bumps and micro-valleys on its surface which bacteria and viruses can hide in. While stainless steel is a welcoming home for microbes, copper will inherently kill them – 24 hours a day, forever. Stainless steel looks reassuringly clean, but it's all just surface lustre.

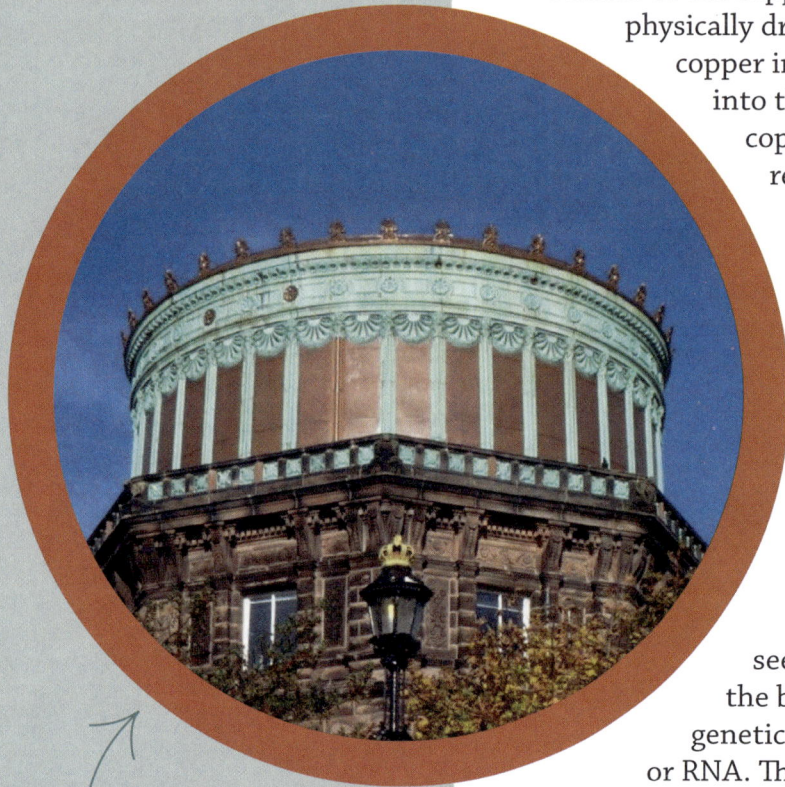

Copper of different ages on the East Tower of the Royal Observatory in Edinburgh, Scotland

MODERN MEDICAL COPPER APPLICATIONS

A recent study looked at the survival times on various surfaces of the virus (SARS-CoV-2) that causes COVID-19.

The virus remained viable for three days on plastic, two days on stainless steel, one day on cardboard – but only four hours on copper. Copper wins hands-down.

Another study looked at hospital-acquired infections, often called health-care associated infections. It examined the use of copper on surfaces in hospital wards, such as bedside rails, intravenous stands or poles, tray tables and chair armrests. This study ran for 43 months, in three hospitals in the USA. The result was that infections in patients dropped by 58%.

Any lowering of infection rates is good, but this is a fabulous reduction. After all, in the USA, hospital-borne infections happen to about 3% of patients, and can cost as much as US$50,000 per patient to remedy. An estimate from 2006 was that patients picked up about 720,000 infections in hospital, leading to 74,000 deaths and some US$125 billion in hospital costs.

Another study concentrated specifically on bed rails in intensive care units. It compared those with plastic surfaces to those with copper surfaces. Over the two-year study, the plastic-surfaced bed rails exceeded the microbial risk standard 90% of the time, while the copper-surfaced bed rails did so only 9% of the time.

Another study in 2016 showed a 78% reduction in drug-resistant microorganisms when the bed rails and the tray tables that slid over the beds had copper-impregnated surfaces. France and Poland are gradually rolling out the use of copper alloys in hospitals.

On one hand, copper is more expensive than stainless steel. But on the other hand, the health costs of treating infections can be enormous. One study (by the Health Economics Consortium at the University of York) claimed the payback time in terms of reduced health costs to the overall hospital budget resulting from the installation of copper surfaces was an astonishingly (and almost unbelievably) low two months.

It seems that we should relearn the lessons of the past, and use copper for touch surfaces (handrails, buttons, taps, doorknobs and push panels) in all public transport, restaurants, kitchens, gyms and buildings. Copper does not need electricity or chemicals to kill microbes – it does this with its own chemical structure, which comes for free.

I've always been a Physics Fanboy, but this Copper Chemistry wins my deep respect. And I'm asking Santa for some copper door handles this year.

Beat egg whites in a copper bowl

Beating the white of an egg can increase its volume by eight times. The liquid egg white is turned into bubbles – which are just small amounts of liquid surrounding air spaces. But besides creating bubbles, the mechanical act of beating the liquid egg white has an effect down at the micro scale – it stretches and uncoils proteins. The theory is complicated, but part of it claims that the foam is stabilised down at the protein level by 'disulphide' bridges, and that copper stimulates their production.

One of the many dozens of known proteins in egg whites is conalbumin. This protein is delicate and tends to fall apart when stretched across the skin of the foam bubbles in freshly beaten egg white. But the copper combines with the conalbumin to make it stronger – so that the protein and the foam don't break down. The combination of copper-and-conalbumin is slightly yellow, giving properly whipped egg whites that classic slightly golden tinge.

One study showed that egg whites that had been whipped in glass bowls were grainy and dry after only 10 minutes. But if they were whipped in a copper bowl, they were still stiff and moist after 20 minutes.

Depending on your budget, you can get your copper from a brass-plated fork, or a copper bowl or whisk.

THE AMAZING DISAPPEARING ANUS

We have recently discovered the very first animal with a disappearing anus! Yep, this animal generates an anus, but only when it needs to poo.

This is really odd – most creatures have an anus that is there 24/7, from birth till death. But a bum that disappears and then reappears, on demand?! Who came up with that idea?

Now, we have to be a little delicate talking about this 'embarrassing' bottom part of the body. It's just too ripe for puns. Yes, talking about it involves a lot of cheek. Is this a crappy area of research? Or do we risk falling behind if we don't do the research? And does doing the research expose us to becoming the butt of others' jokes?

GETTING TO THE BOTTOM OF ANAL EVOLUTION

It turns out that we don't fully understand how the anus came to be – even though it's one of the high points of the last 540-or-so million years of animal evolution. Parasites don't need a digestive tract – but apart from them, practically every animal has one. It starts with the mouth that you pop food into, and continues with the stomach and the intestine, which break down the food into smaller chemicals. Once they leave the stomach, these chemicals are then absorbed into the body of the animal, where they are used both as building blocks and also as nutrients to grow and sustain the body.

But how do animals get rid of what they don't need – the waste?

Well, Nature has evolved two different types of gut to process food – the 'blind gut' and the 'through gut'.

Coral is an animal that has a blind gut (also called a sack gut). It's a simple sack with only one opening. Food comes in through the opening, and then waste products go out through the same opening. Yes, coral eats with its anus.

With the through gut, food comes in via one opening and then passes through a system of tubes to extract nutrients. Finally,

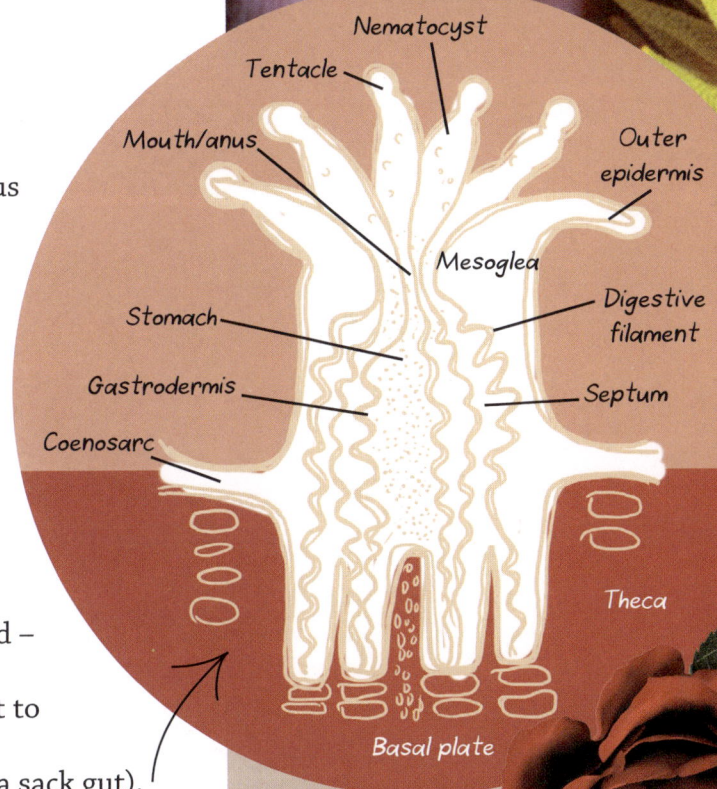

Nematocyst

Tentacle

Mouth/anus

Outer epidermis

Mesoglea

Digestive filament

Stomach

Gastrodermis

Septum

Coenosarc

Theca

Basal plate

Coral polyp

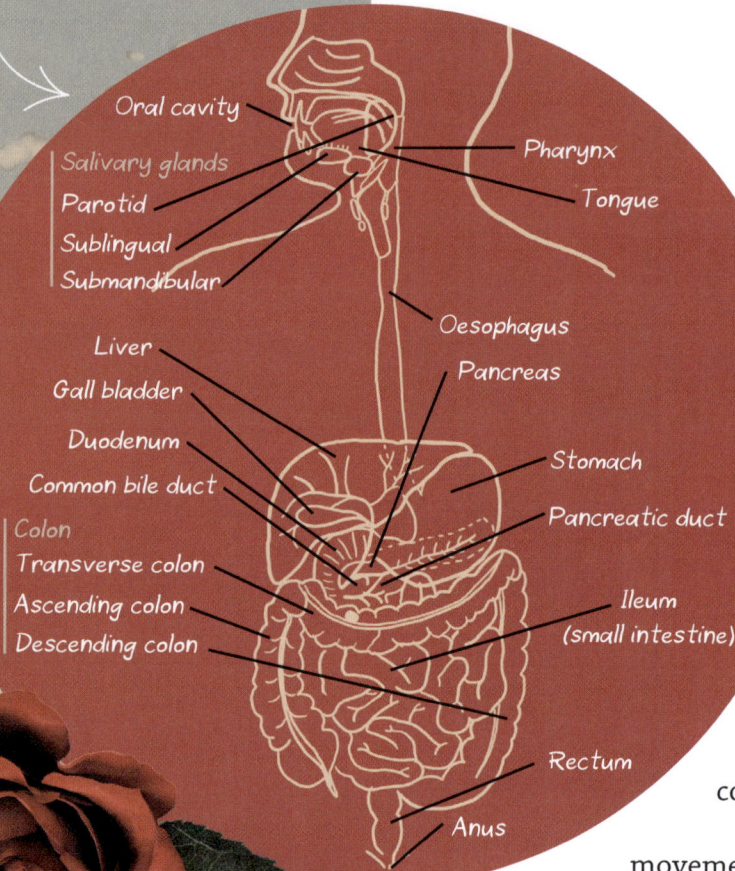

Human gut

Oral cavity

Salivary glands
Parotid
Sublingual
Submandibular

Pharynx
Tongue

Oesophagus
Pancreas

Liver
Gall bladder
Duodenum
Common bile duct

Stomach
Pancreatic duct

Colon
Transverse colon
Ascending colon
Descending colon

Ileum
(small intestine)

Rectum

Anus

A ribbon worm

there's a separate opening in a different location to get rid of the waste products.

The through gut has two advantages. First, you can eat a meal while your previous meal is still being digested somewhere downstream, in another part of your gut. (You can eat lunch, without having to first vomit up your breakfast.) A through gut is a great advantage for the ribbon worm, for example – which is very skinny and very long. The longest recorded ribbon worm was at least 30 metres long (possibly 54 metres long, but they are very hard to measure). Imagine trying to vomit up the entire contents of a 30-metre-long gut.

Second, having an essentially one-way movement of food in your gut means that you can evolve a rather long gut with various specialised areas.

Cows perform the amazing trick of turning grass into meat – even though grass is very low in nutrients. They do this by having one stomach, which is followed by four specialised fermentation chambers along their gut. They can extract the maximum amount of nutrition from plain old grass thanks to the bacteria that have colonised their fermentation chambers. For example, cows get essential vitamin B_{12} from their gut bacteria. If we eat cow meat, we can then absorb this bacteria-produced B_{12}.

As an aside, regardless of the obvious advantages of a through gut, there are several cases in evolution where critters have gone from a through gut 'back' to a blind or sack gut.

It turns out that trying to work out 'the origin of the through gut is mainly a question about the evolution of the anus', according to developmental biologists Andreas Hejnol and José M. Martín-Durán in the paper 'Getting to the bottom of anal evolution'. Yup, if we want to know how our gut evolved, we have to first understand the anus.

Surprisingly, it seems that we still do not have the full answer about the evolution of the anus. However, some recent research has found a few sets of genes that play an important role in anus formation.

TRANSIENT ANUSES – OR A BUM STEER?

Back in 1979, Dr E.B. Knauss, from the University of South Florida, thought a worm might carry a clue to the origin of the anus. It was possible that the gnathostomulid *Haplognathia* could have a transient anus that would appear only for pooing. This critter is a tiny jaw worm, only a few millimetres long. Dr Knauss thought a 'transient anus' was possible, because there was some 'strange' anatomy at the far end of this worm's gut.

In 2002, Drs R.M. Kristensen and P. Funch, from the University of Copenhagen, also thought that the micrognathozoan *Limnognathia* could have a transient anus. This tiny critter was discovered living in the hot springs of Disko Island, in Greenland.

But in each case, the scientists couldn't catch/photograph direct observations of a tiny transient anus, popping into, and out of, existence. At that time, a transient anus had never been caught in the act of being transient.

WARTY COMB JELLY 101

But there is one creature that has been caught in the act. The warty comb jelly has been photographed doing the 'transient anus' thingy.

Evolutionary biologists like to study the warty comb jelly (*Mnemiopsis leidyi*) because it is one of the most ancient forms of complex animals on the planet – about 515 million years old, and it's still here today. They're quite small (up to 10 cm), and belong to the ctenophore (Greek for 'comb') group.

Even though they go back so far in time, they are sophisticated enough to have a separate mouth and anus. They eat tiny crustaceans and baby fish, which gives them the 'fuel' for their 'waste'. They are also hermaphrodites, which means that they can fertilise themselves.

The overall phylum (Ctenophora) has the common name of 'comb jellies', because the groups of cilia (little wavy hairs) that they use for swimming can look like a

Strange anuses

Some creatures, such as birds and reptiles, do not have a specialised anal opening that is dedicated to passing only poo. Instead, they urinate, defecate and reproduce with one single opening, called the cloaca.

Most flatworms do not have an anus. But some flatworms have evolved a single anus. And some flatworms, such as the polyclad flatworm (*Thysanozoon nigropapillosum*) have a whole bunch of anuses (50–100), scattered across their entire back.

The sea cucumber (*Parastichopus tremulus*) has gone multifunctional with its anus. Sure, it uses it to poo – but it also breathes through its anus.

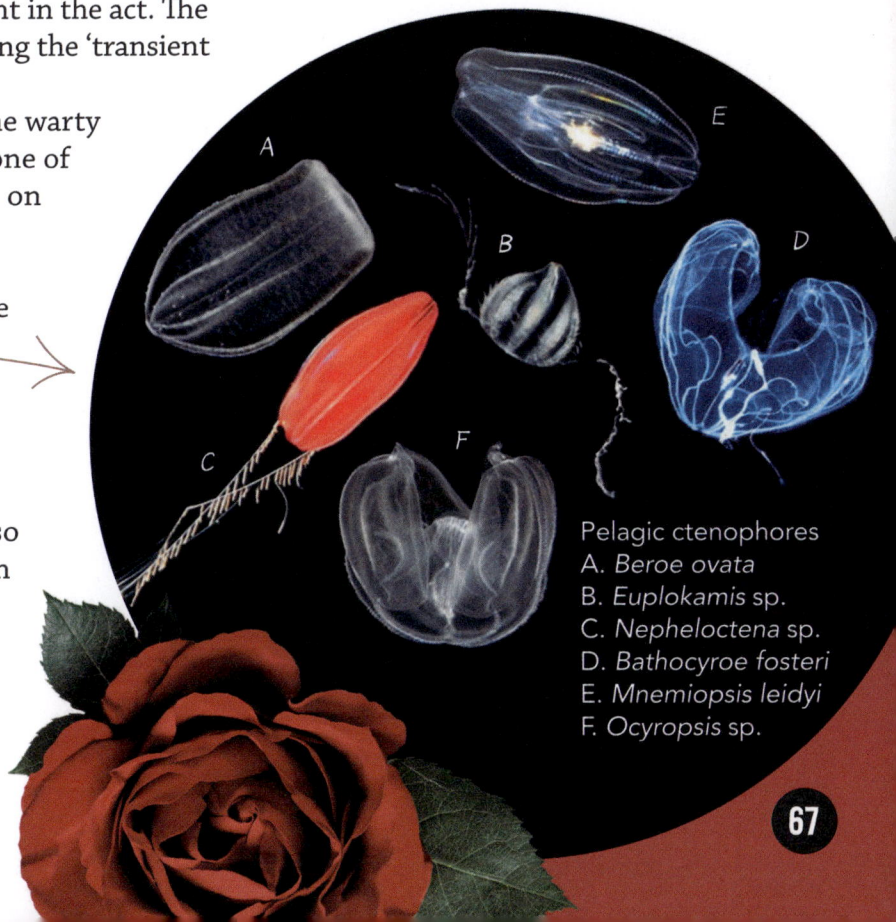

Pelagic ctenophores
A. *Beroe ovata*
B. *Euplokamis* sp.
C. *Nepheloctena* sp.
D. *Bathocyroe fosteri*
E. *Mnemiopsis leidyi*
F. *Ocyropsis* sp.

comb. In fact, they are the largest animals to swim by using cilia. They can range in size from 1 to 1,500 mm long. They are mostly predators – none are vegetarians – and they can eat ten times their own body weight each day. There are about 100–150 different species. They are found almost everywhere – poles to tropics, coastlines to mid-oceans, and waters both shallow and deep. However, they are not found in fresh water. *Mnemiopsis* usually hang around near coastlines.

In the late 1980s, they were accidentally introduced into the Sea of Azov and the Black Sea via the ballast water of ships. They adapted so well, eating fish larvae and small crustaceans, that the size of fish catches dropped precipitously. In the late 1990s, they arrived in the Caspian Sea, the eastern Mediterranean, and are now thriving in the North Sea and the Baltic Sea.

FIRST PROOF OF TRANSIENT ANUS

Way back in 1850, the pioneering zoologist Louis Agassiz saw that the warty comb jelly would expel waste products from its anus. Today we have extra knowledge. We know that it squirts out waste from both ends – the bigger stuff through the mouth, and the small stuff through the anus.

But only in 2018 did Dr Sidney L. Tamm, from the Marine Biological Laboratory at Woods Hole in Massachusetts, actually manage to photograph the expulsion of waste from the anus of the comb jelly.

A. The early stage of defecation from the very swollen left fork of a 5-cm adult *Mnemiopsis leidyi*. There is a very large 'load' in the left anal fork. The arrow shows the beginning of defecation, with the expelling of waste particles.
B. Same view 3.8 minutes later. The left anal fork has shrunk, but is still expelling waste. The arrow shows waste particles still leaving.

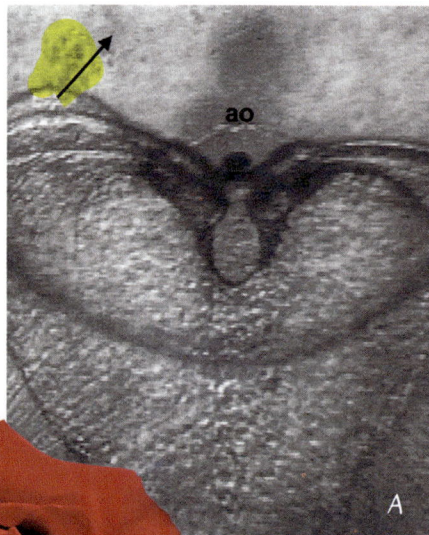

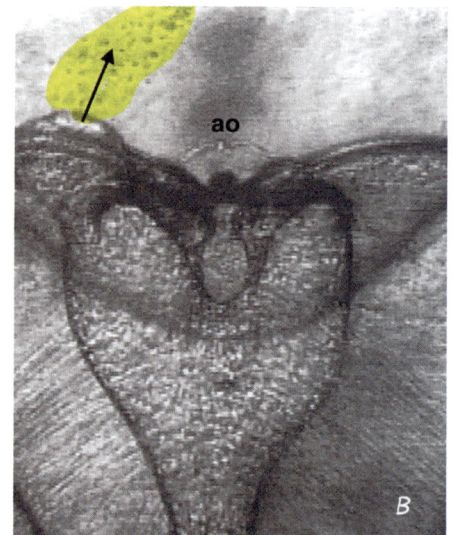

ao

ao

A

B

To his astonishment, he found that 'defecation occurs only through a single anal pore which appears and disappears with a regular … rhythm'. The timing of pooing depends on the size of the creature. If they were small larvae (yup, a junior jelly), it would happen every ten minutes. But large adults, some 5 cm long, would do it every hour or so.

These comb jellies all have an internal right anal fork and an internal left anal fork. These 'forks' are kind of similar to our single rectum – except that there's two of them.

At pretty regular intervals, one of these anal forks begins to bulge mightily, and kiss gently up against the inside of the critter's 'skin'. This newly formed 'bubble' of poo merges with the skin, briefly opens, and suddenly expels its waste products into the outside world. Each comb jelly has its preferred right or left anal fork, using the same one every time. So while we humans are right- or left-handed, the warty comb jelly is right- or left-anused.

And then the 'skin' heals over perfectly, to leave absolutely no hint that the transient anus has ever existed there. The transient anus closes up and vanishes about two to three times faster than when opening up to 'magically' appear.

But there are still so many unanswered questions.

How are the cilia on the inside of the gut involved? Where is the nerve supply that makes the transient anus appear and disappear? Does the anus disappear completely, or is it simply too small to see with current imaging technology? And exactly how does the anus come into existence and then disappear – what tissues are involved? And just for fun, how come the anus is not exactly on the centreline of the body, but off to one side?

But so far, one thing is definite. Dr Tamm said, 'There is no documentation of a transient anus in any other animals that I know of.'

Certainly, our scorpion on page 70 doesn't qualify for a transient anus, because once it's lost its stinger, that's a total bummer!

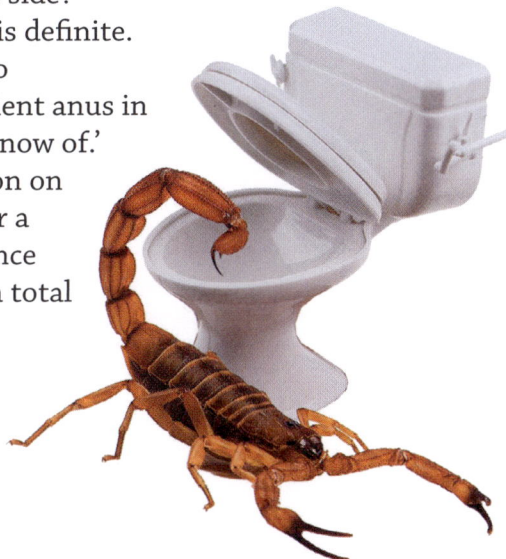

World record for absolute constipation

When I was a very junior medical student at Concord Repatriation General Hospital in Sydney, one of my senior medical bosses in Gastroenterology, Dr Phil Barnes, told us young students a story of the record for 'absolute constipation'. 'Absolute' means that no poo at all had passed through the anus.

Of course, this is an unofficial record. There is no such category in any book of records.

When Dr Barnes was studying in the UK, a medical colleague of his had a young male patient who had not 'passed' anything at all from his bum for six months (not days, not weeks, *months!*). On X-ray, faeces could be seen all the way from the rectum to the diaphragm. The patient's abdomen was quite swollen. Dr Barnes's colleague was a fairly junior doctor at the time, so he was given the job of 'manually disimpacting' the patient. I won't go into details except to say that besides lavage solution and enemas, something similar to a long-handled teaspoon (okay, a parfait spoon) was involved.

Afterwards, the patient was much relieved, of both his discomfort and part of his weight – he ended up about 20 kg lighter.

A medical career won't always smell like a rose garden.

WHERE DID I LEAVE MY ANUS?

Adult male scorpion, *Ananteris balzani*

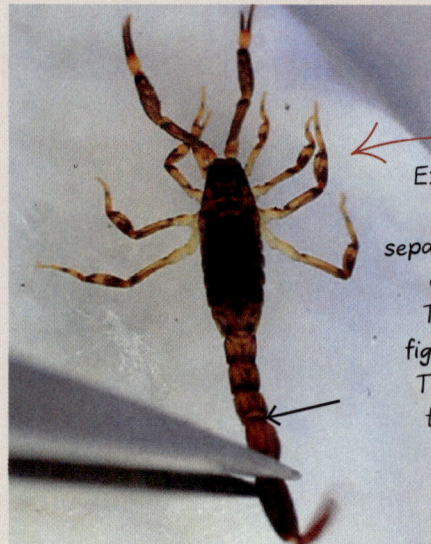

Exact moment before 'tail separation/cleavage' or autotomy. The scorpion is fighting to escape. The arrow shows the beginning of cleavage.

Immediately after the tail has separated. Even though it is fully separated, the detached tail is still twitching.

If you think that losing your smartphone is bad, losing your anus would be worse. Read on – and get ready to pour on the sympathy, because this is a sad tail/tale to tell.

Some scorpions can lose their anus!

Losing-the-anus definitely does not happen to every scorpion – we're talking only about the rare South American scorpions of the *Ananteris* genus. Now, most of us have heard of lizards voluntarily breaking off their tail, and leaving that juicy morsel behind for a predator to eat, so the lizard can scamper away successfully. Sacrificing your tail is a fairly high price to pay, but not as high as losing your life. This self-defence mechanism is called 'autotomy'.

The lizards have evolved to allow autotomy to happen in a fairly painless way. Their tails have several built-in lines of weakness or cleavage, where they can break off

with minimal damage. The wound left behind usually closes quickly and cleanly, with good healing.

Dr Camilo Mattoni from the National University of Córdoba in Argentina looked at *Ananteris* scorpions in the wild. He reckons that about 5–8% of these scorpions have done the 'autotomy' thing with their sting. Yes, instead of a proper sting at their back end, they have a stump.

Now, here's the messy bit – they can also lose their anus.

The anus of this scorpion is not between its back legs. Instead, it's near the very end of the tail, just before the sting. So when these scorpions get attacked, they voluntarily break off some segments of their tail – and unfortunately, in addition to losing their sting, they lose their anus.

You can immediately see a few problems with this. The scorpions can't kill creatures bigger than

themselves with their sting (because it's gone). They now have to survive by eating smaller critters they can catch with their pincers at the front end.

But (and here's the stinger), without an anus, they can't poo.

Even so, they can survive for up

Stages in the healing of the severed stump of adult male *Ananteris solimariae*: A. One hour after separation. B. One day. C. Two days, brown scar developing. D. Three days, scar developing. E. Four days, scar almost fully developed. F. Five days, scar fully formed. G. Ten days, scar darkened. H. Twenty-five days, scar fully defined.

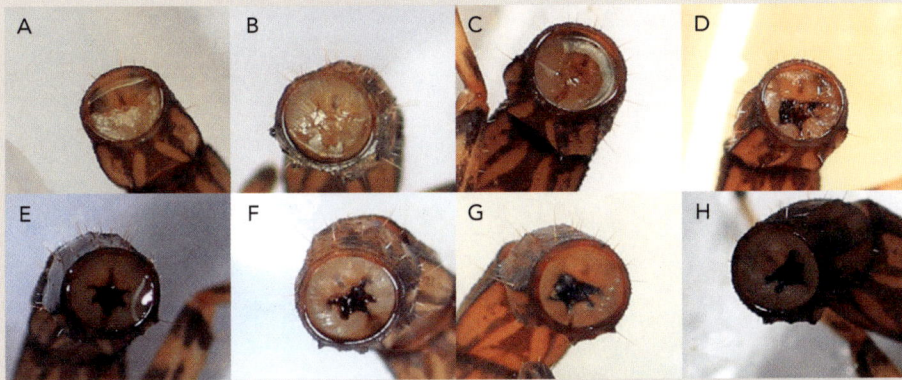

Twenty-five days after autotomy.
A. This scorpion is feeding on a cricket nymph.
B. This scorpion is trying to sting.
C. The arrow points to the accumulated excrement.

to eight months. Usually, because the break is at a segment, it heals well. But sometimes, their waste products build up in their gut and force another segment in their tail to pop off – and release a big load of poo. Even though these scorpions do not have a tail, a sting or an anus, the males can sometimes mate, and father up to 34 babies at a time.

The females are much less likely to voluntarily pop off their tail. First, female scorpions live longer, so from an evolutionary point of view, they have more to lose if they die sooner. Second, they potentially will have baby scorpion embryos to grow and nourish inside themselves. They will definitely need their stings to catch larger critters, and so they avoid autotomy as much as possible – and juveniles can't do it at all.

My second-favourite organ

For a long time, my favourite organ has been the uterus. (Why? It expands from the size of a pear up to the size of a shopping bag and then shrinks down again without any wrinkles or stretch marks, which would be potential sources of weakness in a subsequent pregnancy when the uterus must expand again – and because I can call it a 'uter-house', a home for growing a baby 😊.)

But for me, the anal sphincter comes in at Number 2 (get it?) – because of how well it handles farts.

Here's a little experiment for you. Interlock the fingers of your right and left hands, and make a neat temporary cup for a bit of water. Add a few little solid lumps to the water. And remember that you've got the air (which is a gas) above the water.

Now comes the catch. Can you do any action with your hands that allows the gas to pass downwards through your interlocked fingers, while still keeping the solids and the liquids above?

Nope, you can't. But your anal sphincter is able to do something very similar! Even better, it does this several times each day – usually with no accidents (once you're not a baby anymore).

Hooray for the anal sphincter and continence!

MURDER HORNETS –
LETHAL BUT TASTY?

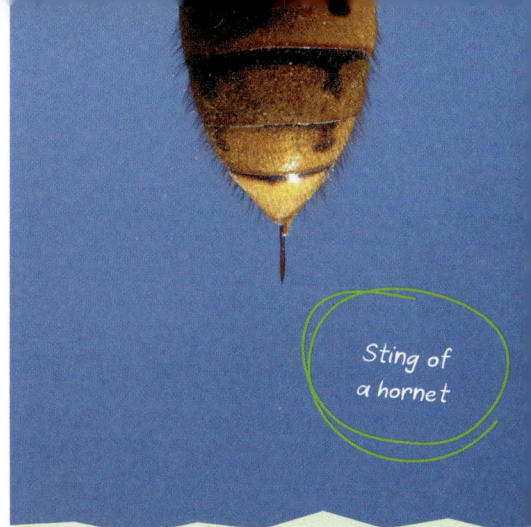

Sting of a hornet

The New York Times headline promised so much. 'Murder hornets: Both a lethal threat and a tasty treat in Japan'. It made me want to read more – as was the intent of the subeditors – and it succeeded!

(By the way, the entomologists – the insect scientists – get quite annoyed by the media nickname of 'murder hornets'. Quite rightly, they see it as unnecessary media hype by TV variety shows and tabloid news outlets. But the name got me reading and learning about them, so …)

ASIAN GIANT HORNET 101

Actual size

Its formal name, *Vespa mandarinia*, sounds quite sweet – a mix of an Italian motor scooter and a small citrus fruit. However, this stinging insect kills a dozen people each year – or maybe more.

The Asian giant hornet is also called the Japanese giant hornet. As advertised by its name, it's the largest hornet in the world. It's about 45 mm long and has a wingspan of around 75 mm. Clothes don't protect you from its sting – its 6 mm-long stinger can penetrate three layers of fabric.

Asian giant hornets live in colonies, often underground. They are similar to honeybees in that each colony has a single fertile queen, many sterile female workers and a few males – their only duty is to fertilise the queen, so they are booted out when their work is done. The males don't have a stinger.

These hornets are pretty specific about where they live. They don't like plains or high-altitude climates – they prefer low mountains and forests. They can fly up to 100 km in a day, at speeds of up to 40 kph.

They are most commonly found in Japan, but they also live in China, Korea, the Russian Far East, India, Nepal, Myanmar, Laos, Malaysia and Taiwan.

In late 2019, a few colonies were found in northwestern USA and southwestern Canada. The worry is that they might attack honeybees there.

A defensive ball of Japanese honeybees (*Apis cerana japonica*) in which two hornets (*Vespa simillima xanthoptera*) are being engulfed, incapacitated, heated and eventually killed. This sort of defence is also used against the Asian giant hornet.

HORNET INVASION – A KILLING SPREE

Honeybees are already under threat from a whole lot of causes, especially colony collapse disorder (a phenomenon that has caused a drastic decline in bee populations around the world) – they definitely don't need any more enemies.

In the USA, since 2012, the honeybee numbers in hives have been declining by 29–45% each year. The Winter of 2018–19 was the worst on record for American honeybees. And yet we really need honeybees.

In the USA, billions of honeybees carry out an essential, but unpaid, job. They pollinate more than 90 agricultural crops. This contributes about US$15 billion each year to the US economy.

When it comes to honeybees, the Asian giant hornet is a brutally ruthless predator. A single coordinated attack by a few dozen hornets can wipe out the many thousands of bees in a hive in just a few hours. A single patrolling female hornet will not usually attempt to attack a beehive by herself. Instead, she will mark the beehive with a pheromone, and then return with between 2 and 50 colleagues – classic bully tactics.

The hornets are some four-to-five times bigger than a bee. They don't bother stinging the bees, because that would use up their venom – and there are thousands of bees to kill in the next few hours. Instead they use their large mandibles to *decapitate* the bees, a scenario which is as gory as a Tarantino film. (By the way, a honeybee dies after a single sting when its barbed stinger gets pulled out of its abdomen as you brush the bee off your skin. But an Asian giant hornet can sting many times without any damage, because it has a smooth stinger which can be inserted and extracted easily with each sting.)

In one study, scientists counted hornets, on average, killing one bee every 14 seconds. However, there have been some reports of much faster killing rates. The entomologists bluntly call this the 'slaughter and occupation phase'.

BEES FIGHT BACK WITH HEAT AND GAS

There are more than 30 species of honeybees endemic to Eurasia and northern Africa. A few of them (such as the Japanese honeybee *Apis cerana japonica*) have worked out an effective defence against the Asian giant hornet.

If a honeybee is lucky enough to detect a scouting hornet as she enters the beehive, the bees have a chance. The honeybee can emit a pheromone, which attracts 100 or so other female honeybee workers.

There is no point in trying to sting the hornet, as she is protected by a rigid exoskeleton, which the stingers of the honeybees can't penetrate. So, several hundred honeybees deploy a better counterattack mechanism. They rush at the hornet and try to totally wrap themselves around her – they turn themselves into a ball of bees. As they engulf the hornet, they violently vibrate their flying muscles. This is hard work, so they generate lots of carbon dioxide, and lots of heat – temperatures of up to 46–50°C. The vast majority of the honeybees will survive the high temperature and the high carbon-dioxide level. But the Asian giant hornet usually cannot, and she dies.

Basically, the honeybees cook the living hornet to death! This is probably fair enough if the only other option is losing your own head.

This tactic works only if there are just a few hornets invading the hive. It is a very 'expensive' fight for all the parties involved; if there is just one hornet involved, she dies, and so do some honeybees. But if 50 hornets descend on the beehive at the same time, the defending bees can't physically get close enough to envelop all the invaders with their bodies. The honeybees are goners.

Some honeybees have developed another strategy. They shake their little honeybee body in a specific pattern. This sends the hornet a specific signal – not a threat signal, but an 'I see you and I have lots of friends' signal. In many cases, the hornet simply backs off, and looks for a more vulnerable target.

Western honeybees (*Apis mellifera*) have been introduced into Asia. They have the advantage of producing lots of honey – and being quite gentle or tame. Unfortunately, not having evolved with the Asian giant hornet, they have no defences against it.

Pheromones

A pheromone is like a hormone, except that it travels outside the body to another creature, usually of the same species – and it can be as powerful as a hormone in how it can change behaviour.

Nature uses pheromones for three very important jobs – food, sex and death.

Food? If a foraging ant stumbles across some food (like your picnic), it will take a sample back to the nest, but at the same time, squirt some pheromone onto the ground as it travels. Any other ant that stumbles across that pheromone trail will immediately head for your picnic.

Sex? A female gypsy moth, when she is ready to mate, will release some pheromone into the air. Soon, males will come a-courting.

Death? If you are an ant, or a wolf, you have to smell like an ant or a wolf – and especially from the right group. Yes, gang colours count. If you smell different, other group members will kill you (nothing personal, it's just pheromones). Honeybees and hornets also use pheromones for communication.

Male gypsy moth

75

Tasty treat

In the central Chubu region of Japan, the Asian giant hornet is used as an invigorating ingredient for alcoholic drinks, as well as a nutritious snack. It's even claimed to be an aphrodisiac.

The hornets can be eaten as grubs – pan-fried or steamed with rice. The bigger adults are fried on skewers, and then eaten complete with the stinger and the venom. Not surprisingly, they have a special taste sensation – warming and tingling. In Tokyo, the hornet is served in some 30 restaurants.

The venom is also used to give alcohol an extra kick. The live hornets are drowned in a clear distilled alcohol called 'shochu'. As they die, the hornets squirt out their venom. The liquid then sits until the chemical reactions between the alcohol and venom turn it a dark amber colour. Another use for the processed venom is as a supposed arthritis cure.

MURDER HORNET – HOMICIDAL OR JUST DEADLY?

Being called a 'murderer' is pretty damning, even if you are a hornet. So is it fair?

The venom of the giant Asian hornet is not particularly potent, but there is a lot of it – seven times more than in a honeybee (1,100 micrograms vs 150 micrograms). That's enough venom to have a 50% chance of killing a 270-gram rodent. Each year, one or two dozen people die from a giant Asian hornet sting in Asia – but that's from multiple stings.

In Japan alone, since 2001, the number of people killed annually by the stings of bees, wasps and hornets has ranged between 12 and 26. We have an overall tally, but we don't have the exact number for Japanese deaths caused by the Asian giant hornet alone.

On average, people who died due to the Asian giant hornet had been stung 59 times, while the survivors had been stung 'only' 28 times. One American apiarist got stung by an Asian giant hornet seven times, and said each sting was like 'being pierced by a red-hot nail'.

In 2013, in the Chinese province of Shaanxi, the Asian giant hornet killed 41 people and injured more than 1,600.

The official advice in China is that if you get stung more than ten times, you should seek medical assistance. But if you get stung more than 30 times, you should go immediately to the emergency department. One side effect of the venom is kidney failure – and you need working kidneys to stay alive.

Every Spring, the Japanese government issues special advisory notices about the Asian giant hornet, which the Japanese call 'osuzumebachi', which translates as 'giant sparrow hornets'. (Actually, they are not as big as a sparrow.) Perfumes and hair spray seem to attract the hornets, so people are advised against getting 'dolled up' when they go bush.

So these hornets aren't exactly out there deliberately slaying thousands of us every night. How then did they get the sensationalist nickname 'murder hornet'?

It happened as a result of a slight mistranslation. In the Japanese popular press, they are referred to as 'satsujin bachi', which is simply 'killer hornet'.

They were never called 'murder hornet' in Japan. After all, 'murder' involves some degree of pre-meditation or planning.

So we Australians might talk about a 'killer shark', but not a 'murder shark'.

In Japanese, the word 'satsujin' is written with the characters 'kill' and 'person'. But if you, as a non-Japanese speaker, look up 'satsujin' in a Japanese-English dictionary, you will find many options – 'killer', 'manslaughter' and, yes, 'murder'.

Who can blame the gutter press if they go for the most attention-grabbing translation of 'satsujin' when they discuss this insect?

MAJOR THREAT?

The hornet almost certainly entered the USA in a shipping container. The very real concern about the 'murder hornet' in the USA is that it might get a firm foothold. Dr Chris Looney, an entomologist at the Washington State Department of Agriculture, said, 'This is our window to stop it from establishing. If we can't do it in the next couple of years, it probably can't be done.'

Dr Looney also debunked their scary moniker: 'They are not "murder hornets". They are just hornets.'

By comparison, mosquitoes kill millions of humans every year, by infecting us with malaria, dengue fever, etc. Asian giant hornets? They don't even get into triple figures – they're not even trying.

So the 'murder hornet' is not a major player in the Death League Tables – but it's still good to keep them out of new areas (and especially beehives), if we can.

Anti-hornet measures

One device used in Japan is a special trap that is placed in front of the beehive. It tries to catch the hornet before she enters.

Another possibility is to install doors on the beehives that are too small for a hornet to enter, but big enough for the honeybees to pass freely.

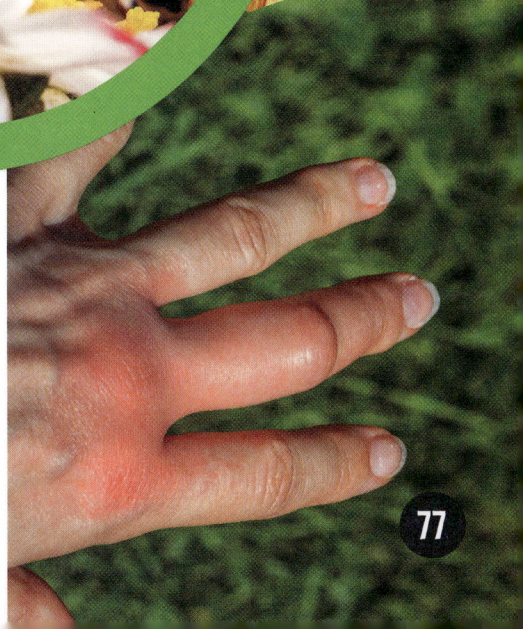

WOMEN'S WORK
— NEVER DONE, NEVER PAID

In Iceland, on 24 October 1975, the economy crashed to a grinding halt. The reason? It was simple: 90% of the women in Iceland refused to work – paid or unpaid.

For that entire day, Icelandic women did not show up for their jobs outside the home – for which, on average, they received 60% of the male salary. They also did not perform any unpaid labour in the home, such as cooking meals, cleaning the house or looking after children.

The Organisation for Economic Co-operation and Development (OECD) defines 'unpaid labour' to include caring for household or non-household members, shopping for essential household goods, and household maintenance, such as the housework. In general, unpaid labour is totally invisible to economists.

But on 24 October 1975, Icelandic women took a stand, making their unpaid labour suddenly very visible. The Icelandic men compensated as best they could. In other words, they headed to restaurants in droves. And all the workplaces were running wild with children.

This day became immortalised as the 'Long Friday', for reasons women were already familiar with. Suddenly, men had found how long a day could be.

GLOBAL ECONOMICS

So what's changed since that dramatic demonstration?

Worldwide, the unpaid labour of women today adds up to over 12% of the global economy. In 2019, it was worth about US$11 trillion. That is a huge amount – roughly three times greater than the wealth generated by all of the tech companies on the planet, such as Apple, Google, Facebook and so on.

At the very top of the world's economy, we have billionaires – about 2,153 of them in 2019. Between them, their wealth is greater than the combined wealth of the poorest 4.6 billion people on the planet. That's about 60% of the world's population, which was about 7.8 billion in March 2020.

But at the bottom of the world's economy, every day, women and girls are working 12.5 billion hours for free, in traditional caring roles. That's equivalent to 1.5 billion people working a full eight-hour day for nothing – and doing this every day of the year.

Care is not optional – it is both a critical social good and also an essential human right. But it is overwhelmingly done by underpaid or unpaid women and girls. Care workers are paid less than workers in fields that need similar levels of skill and qualifications.

Roughly 1% of the world's population (about 67 million people) are domestic workers. About 80% of all domestic workers are women. The vast majority of them have no access to maternity leave or health insurance – which most workers would consider the basic perks of being employed. About 50% of domestic workers have no minimum wage and no legal limit on their work hours.

FEMALE POVERTY

The World Bank defines 'extreme poverty' as having an income of less than US$1.90 each day. Close to 10% of the world's population (735 million people) live in extreme poverty. Part of living in extreme poverty is the day-to-day reality of going hungry.

The economic system is stacked against women and girls who are also poor. Young girls who do lots of unpaid care work spend less time at school than other girls. This education gap widens after adolescence, because their fewer years of schooling reduces their opportunity for post-secondary education. This has a huge negative impact on their income prospects.

Flowing on from this bad start, the poverty gap between men and women increases further during women's peak productive (employment) and reproductive (child-bearing) years.

Poor women are the poorest of the poor.

THE WORLD FAIR – LET'S MAKE IT SO

But a fairer world *is* easily achievable.

In a fairer world, every child can fulfil their potential, all people can have secure jobs paying decent wages, and people do not have to live in fear of falling sick or a failed harvest.

There's enough money in the global economy to do this –

it just needs to be redistributed. Even though some 2,153 billionaires own 40% of the wealth, they pay very little tax. The multi-billionaire Warren Buffett said that he pays a lower tax rate than his secretary. Billionaires benefit from highly paid accountants, favourable taxation laws and the ability to easily shift huge amounts of money across the world into zero-tax zones. Their average annual increase in wealth over the last ten years is 7.4% – much greater than bank interest. Even just a tiny 0.5% extra tax on the wealth of the richest 1% of the world would be enough to create 117 million jobs. Think what this could do for education, health and aged care, etc.

Governments can 'govern' to fix this economic injustice. In 2020, Oxfam released the report 'Time to care: Unpaid and underpaid care work and the global inequality crisis'. They recommended investment in national care systems (child care, aged care, etc) to relieve the extra caring responsibility from women and girls – and much more.

The coronavirus pandemic showed that governments *can* stump up and spend the dollars when they see (especially in the short term) the high economic and social costs of doing nothing. If they can see the long-term benefits of investing in reliable universal care and setting a basic minimum wage for all, regardless of gender, we can begin to make a world where half its population isn't unfairly disadvantaged merely because they carry an additional X chromosome.

So, what was the result of the 'Long Friday' of 24 October 1975, when the women refused to do labour of any kind? A year later, the Icelandic government passed laws guaranteeing equal pay for women in the workforce. Although unpaid labour has still not been addressed, it was an important win for all of Iceland – not just its women and girls. Iceland is one of the most progressive countries in the world in terms of women's equity – and it tops the Global Gender Gap Index, in terms of equality.

A Chinese saying goes, 'Women hold up half the sky.' It would be good if they got paid for it.

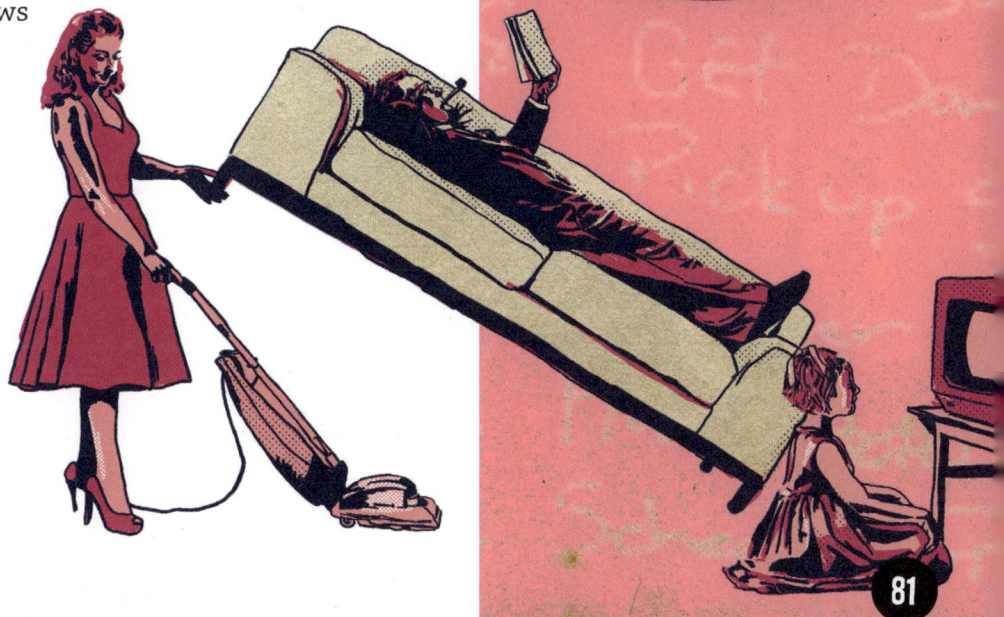

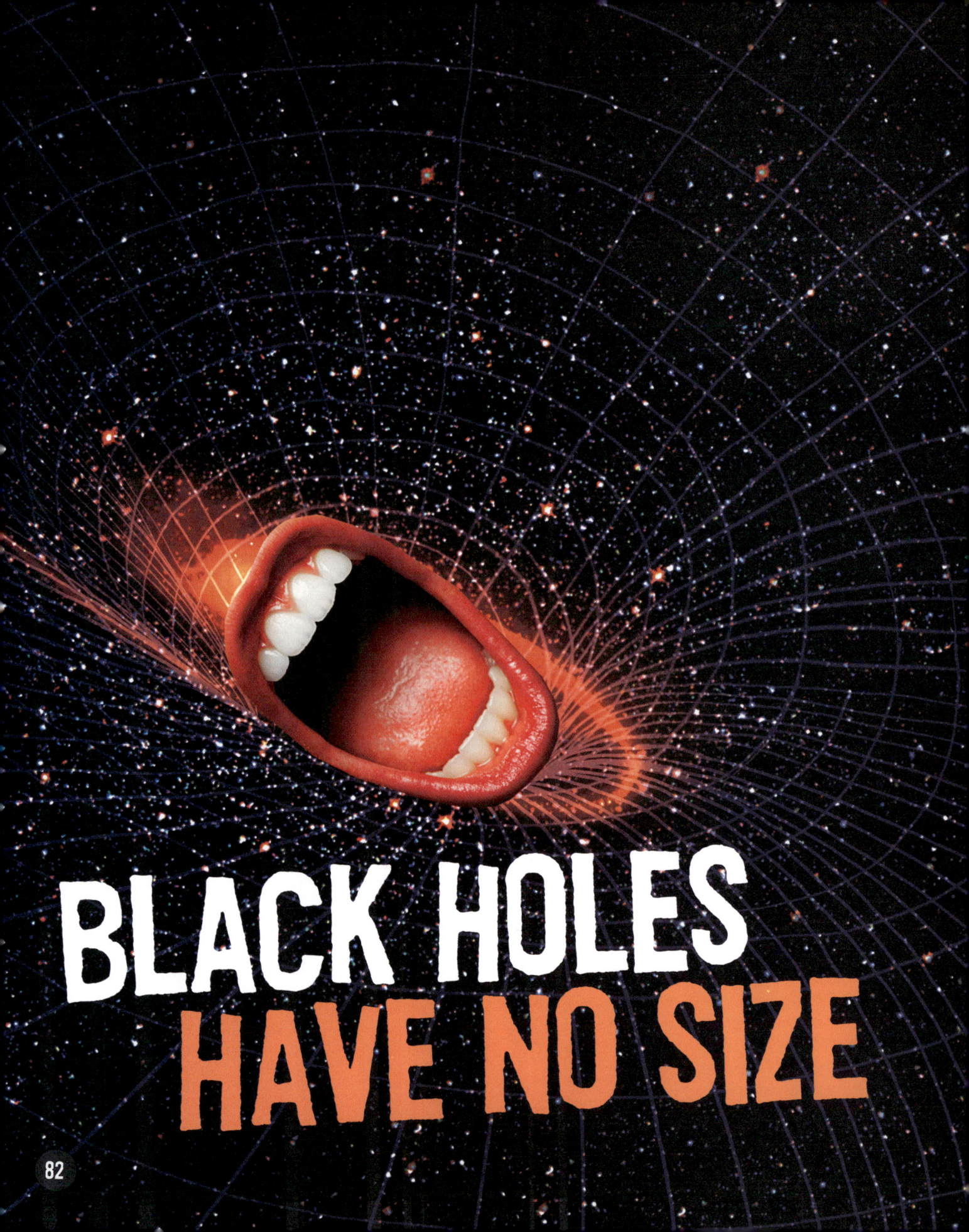

BLACK HOLES
HAVE NO SIZE

Get ready for some mind-blowing black hole sciencey goodness.

Every black hole is exactly the same size. That size is zero. It doesn't matter how massive the black hole is, it still has no size.

A black hole is just a point in space where a lot of matter is gathered up into one location. The size of that location is not small, or tiny, or really, really microscopic – it's zero, zip, zilch.

ZERO SIZE, INFINITE DENSITY

Black holes have only three properties – mass, charge and angular momentum (spin).

'Size' is not a property of a black hole, in the same way that being the World Champion Heavyweight Cage Fighter is not one of my properties.

We measure the mass of black holes in Solar Masses. A Solar Mass equals the mass of our Sun – which is about 2 billion billion billion tonnes, or about 333,000 times the mass of the Earth.

But mass is totally irrelevant to size.

It doesn't matter if a black hole is 3 Solar Masses (which is about as low in mass as black holes can theoretically be).

It also doesn't matter if it's 4.3 million Solar Masses (like Sgr A* – pronounced 'Sagittarius A-Star'– which sits in the centre of our Milky Way galaxy, some 26,000 light years from Earth).

It's the same for a black hole with a mass of 6.5 billion Solar Masses (like the supermassive black hole at the core of the supergiant elliptical galaxy Messier 87, some 53 million light years away).

All black holes are the same size – which is zero! In other words, every black hole is just a point in space. (I just have to add a rather disturbing subtlety here. Suppose the black hole is rotating. In this case, the 'singularity' – a word physicists use to mean that the Laws of Physics don't apply in this location or situation – is not a point, but a ring, but the volume is still zero. My head hurts – but in a subtle way ...)

Now, a huge amount of mass stored in zero volume inevitably leads to 'stranger things'.

A black hole is so dense that nothing close to it can escape its gravitational pull.

Think about two celestial objects with the same mass – say,

GREEN ROOM

Black holes – not dumb, just really dense

Density means the mass of an object divided by its volume. The density of water is 1 kilogram/litre:

Density = Mass/Volume

The black holes we have measured so far have masses ranging from about 3.8 Solar Masses, through millions of Solar Masses, right up to many billions of Solar Masses.

Okay, so we know the masses of these black holes.

To mathematically work out the density, we need to know the volume. That's very easy. If their size (i.e. their diameter) is zero, then their volume is also zero.

In primary school, you might have tried dividing a number by zero. And you would have been told either it was impossible, or that the answer was 'infinite'.

That's right, the density of a black hole is infinite – at the location of the singularity of the black hole.

5 Solar Masses. The big difference is that with a black hole all that mass is concentrated into a single point, with no volume. But with a star, that same mass is spread out in a ball a few million kilometres across.

As you get closer to the black hole, the gravitational field increases very rapidly – because all that mass is concentrated inside a single point (which has zero size). But as you get closer to a star, the gravitational field increases slowly – because the same amount of mass is smeared across a large volume of space.

But what happens when you get a long distance away from the black hole, or the star with the same mass? At great distances, the gravitational field is the same for both.

With a black hole, the boundary at which things get 'really weird' is called the 'event horizon'.

ESCAPE VELOCITY 101

Within the boundary of the event horizon, nothing (neither matter nor radiation) can ever leave. Why? Because the 'escape velocity' (explained below) inside the event horizon is greater than the speed of light. (The reason nothing can escape is that, in general, nothing can travel faster than light.)

One useful way to think about the event horizon is that it's at the distance from the black hole where the escape velocity from the black hole is equal to the speed of light.

Now, that is a huge amount of information to shove into a single sentence, so let's break it down.

What is 'escape velocity'? It's the velocity, or speed, that an object needs in order to leave the surface of a planet or star or moon and not be pulled back down by gravity.

To make it easy, let's look at the Earth and pretend that there is no atmosphere (and therefore no wind resistance), and that the planet is not spinning.

Stand on the Earth's surface, and hit a tennis ball upwards with a bat or racquet. It will go up some 10–20 metres, and then come down.

Thrust it up at 1 km/second. It will take longer, but after a while, the ball will come back down.

Same for 10 km/s – now it will take a lot longer, but it will eventually come down again.

But everything changes once you can throw the ball at 11.186 km/s. It will escape the gravity of the Earth – forever. It will never return to Earth (unless you chase it with a spaceship, and bring it back).

So, escape velocity is the velocity an object needs to totally escape the gravitational pull of a body (when it's leaving from the surface of that body).

The escape velocity for an object to forever escape the gravitational pull of the Moon is about 2.38 km/s; for Jupiter, about 60.20 km/s; and for the Sun, about 617.5 km/s.

The very first spacecraft we humans were able to give escape velocity to was the Soviet *Luna 1* satellite, back in 1959. But that escape velocity was only for escaping the Earth. It certainly did not achieve escape velocity from the Sun, and so it's still orbiting the Sun.

So, to repeat, the event horizon around a black hole is where the escape velocity is equal to the speed of light.

EVENT HORIZON 101

If you are heading towards a black hole and have crossed the event horizon, you cannot escape.

But if you are outside the event horizon, and you have enough propulsion, with a big enough engine you could avoid being sucked into a black hole. (This is in keeping with the unofficial motto of the US Air Force, 'With enough energy, a pig will fly.')

So here's a weird, but accurate, scenario. Suppose that you are in a rocket hovering just above the event horizon. Even though the rocket is accelerating away from the black hole at a colossal rate, you won't be moving, because the acceleration caused by the firing of the rocket's engines will be exactly equal to the gravitational pull of the black hole. You would, however, be squashed against the back wall of the rocket due to the rocket's powerful acceleration.

Light cannot escape from inside the event horizon. This means that if you're outside the event horizon, you can never see any event that happens inside the event horizon – which is how it got its name.

Let's start looking at the size of various event horizons (which is what the media really means when they talk about the size of a black hole).

Suppose the Sun got turned into a black hole. (Even if all the nuclear fires went out in the Sun, it couldn't collapse into a black hole, because of something called 'electron degeneracy pressure'. But let's suppose we have an advanced technology that can overcome this 'degeneracy pressure'. Check it out in Wikipedia.) The event horizon surrounding our 'black hole/former Sun' would be about 3 km away from the black hole.

Luna 1, launched in 1959, was the first man-made object to attain escape velocity from Earth, but we could not give it enough velocity to escape from the Sun.

DR KARL'S Q+A

Do all galaxies have black holes at their centre?

It does seem that all galaxies have a massive black hole at their centre – and we don't know why.

The diagram on page 89 shows a black hole and its accretion disc, located at the centre of a galaxy. Because the black hole is eating stuff, it is active (e.g. beaming out relativistic jets), and so the galaxy is called an active galaxy.

Our Milky Way is a spiral galaxy, and this shape does superficially resemble the accretion disc that forms around a black hole. However, our entire galaxy itself is not an accretion disc that our Solar System has been pulled on to. But there is an accretion disc around the black hole at the centre of our galaxy – Sgr A*.

Here's a list of celestial objects, their masses and associated event horizon radii. The first two are not black holes, but for the purpose of this exercise, let's pretend that they've been turned into black holes.

Object	Solar Masses	Radius, Event Horizon
Earth	0.000003	8.9 mm
Sun	1	2.95 km
Sgr A*	4.3 million	12.7 million km (¼ x distance Sun to Mercury)
Messier 87	6.5 billion	19 billion km (5 x distance Sun to Neptune)

IT'S HARD TO IMAGINE ...

There's often confusion among non-astrophysicists about the difference between the size of a black hole and the event horizon. In 2019, for the first time, scientists managed to complete two years of very complex work to obtain a single photo of very energetic activity happening around an event horizon that, in turn, was surrounding a black hole.

The galaxy Messier 87 was discovered in 1781 by Charles Messier, a French astronomer. It's one of the most massive galaxies in our patch of the Universe, and it's some 53 million light years away. While our Milky Way galaxy has 150–200 globular clusters of stars orbiting around its outskirts, M87 has about 12,000 – that's a lot more. At the galaxy's centre is the supermassive black hole. Each year, 2–3 Solar Masses of gas fall in towards the centre of M87. A huge jet of superheated 'gas', some 5,000 light years long, squirts out, at close to the speed of light, from the accretion disc – the structure formed by gas, dust and matter that swirls around the central black hole. This jet is enormous. For comparison, the nearest star to us (beyond our own Sun) is just 4 light years away. The energy in the jet alone is about 10 trillion times all the energy produced by every star in our Milky Way galaxy in each second. The accretion disc is rotating at about 1,000 km/s, is about 0.4 light years across, and is a long way outside the event horizon.

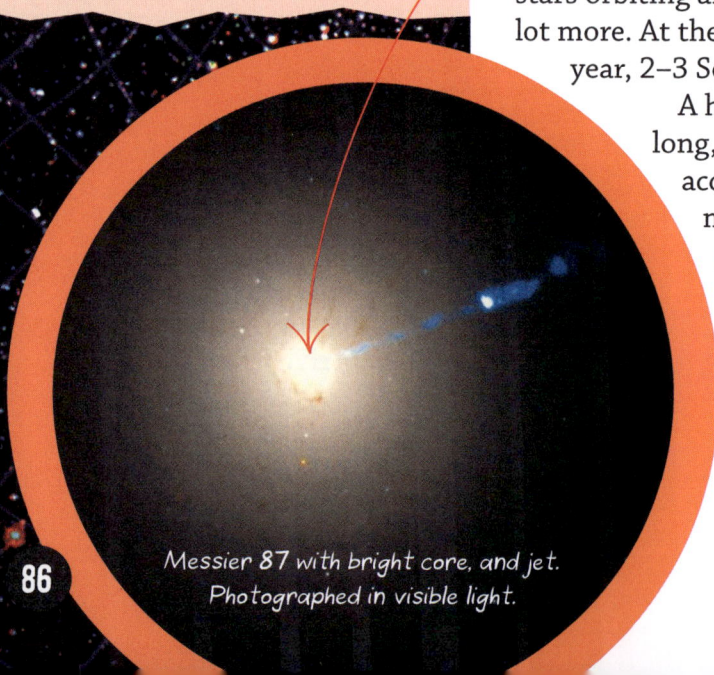

Messier 87 with bright core, and jet. Photographed in visible light.

The enormous astrophysical jet squirting out of the accretion disc of M87, photographed in radio frequencies, not visible light. It's 5,000 light years long.

You get the idea – this is a very big galaxy, with lots of energetic stuff happening around the 6.5 billion-Solar-Mass black hole.

The single photograph (on the next page) took a huge effort to make. The name for this project was Event Horizon Telescope (EHT). Hundreds of astronomers from some 80 different institutions had to collect observations from half a dozen sets of radio telescopes scattered across the surface of the Earth – some at the South Pole. The collected data (5,000 terabytes) was stored on half-a-tonne of hard drives, and then processed by four separate teams, each using different algorithms and mathematics. The resulting four single pictures were remarkably close to each other. They were also very close to what the astronomers simulated on computers. This gave them confidence that they were on the right track. And finally, they released it – two years of work to get one single photo!

After all this meticulous hard work, the popular media reports only managed to get the first sentence right, but the next two sentences were wrong (sob).

'New EHT measurements show that the mass of this black hole is about 6.5 billion solar masses.' Correct.

'The team also has figured out the behemoth's size. Its diameter stretches 38 billion km (24 billion miles).' Wrong. The 'behemoth' does not have a size of 38 billion km! That number was the diameter of the event horizon. Black holes have no size, remember?

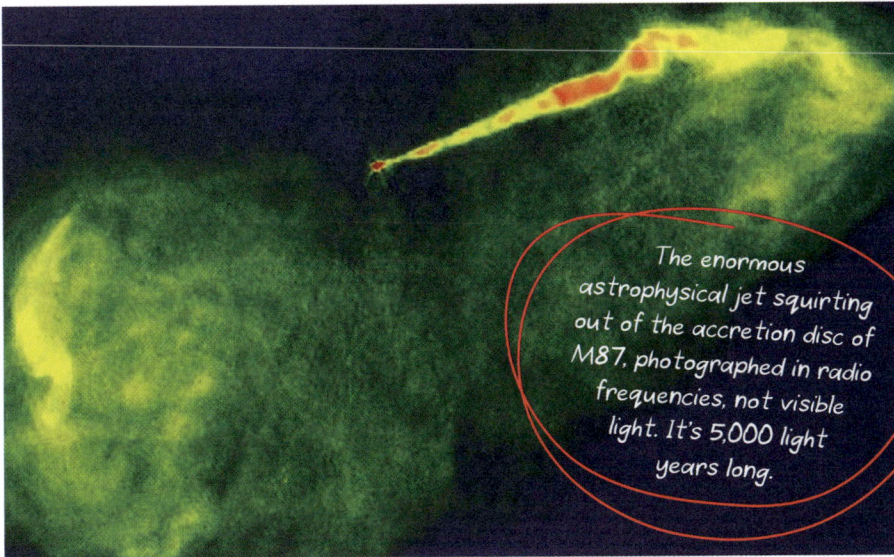

But if black holes seem all too complicated, all you have to remember is this: 'one size fits all'!

The black hole as vacuum cleaner?

A common myth about black holes is that they are cosmic vacuum cleaners. They supposedly suck up all the stuff near them, and in fact, everything out to the edge of the Universe (if it does have an edge – but let's leave that for another time).

Not true.

Way outside a black hole's event horizon, the 'suck' of its gravity is the same as a star with the same mass as the black hole. The big increase in 'suck' is only when you get close to the black hole. Close in, the black hole's gravitational field is a lot stronger than a star with the same mass – because all that mass is concentrated into a volume of zero, rather than being spread out.

If our Sun were magically turned into a black hole (something to do with 'degeneracy pressure'), would it gobble up the Earth?

Nope.

The Earth and all the other planets (from Mercury to Neptune) would continue to orbit in exactly the same orbits. The only difference would be that no radiation of any sort would leave the Sun. So the planets would very rapidly start to cool down, without their daily dose of heat.

Spaghettification

'Spaghettification' (stretching like spaghetti) is what happens to objects as they fall into (or approach close to) a black hole – they get longer and thinner.

Stephen Hawking described the fate of a hypothetical astronaut falling feet first into a black hole. Because all the mass in a black hole is stuffed into a volume of zero, the gravitational field increases quite quickly as the astronaut gets closer. At some stage, gravity's pull on their feet will be greater than its pull on their head, and they will get 'stretched like spaghetti', according to Hawking. The location where this happens depends on the mass of the black hole.

With a black hole of 10 Solar Masses, the event horizon is about 30 km away from the black hole. The astronaut would be 'spaghettified' many hundreds of kilometres outside the event horizon.

But for a black hole of 10,000 Solar Masses, the event horizon is about 30,000 km away from the black hole. The physics tells us that the gravitational field would increase only slowly as you get close to the event horizon. So our astronaut would not get 'spaghettified' as they passed through this event horizon.

FIRST PIC OF A BLACK HOLE SHADOW

This looks like a fuzzy out-of-focus photo of a coffee stain. It's not – but it did take two years and eight telescopes to make.

It is not a photograph of a black hole – after all, they are completely dark, and nothing (including light) can escape them. And it's not a photo of the event horizon – that's just a boundary in space, surrounding a black hole, where the escape velocity of anything fired radially away from the black hole is equal to the speed of light. (The distance of the event horizon from the black hole is called a Schwarzchild Radius, or rs.)

However, this is a photo of the 'shadow' of a black hole, surrounded by a ring of coloured light from the millions of stars on the other side of the black hole.

First, some more about black holes. At about 3 rs is the inner part of the 'stable' accretion disc. (Anything inside this stable inner part of the accretion disc will very quickly end up inside the black hole.)

Now here's something really weird: the Photon Sphere, which is a thin shell of photons. Light can orbit around the black hole, inside this Photon Sphere, at about 1.5 rs. (Light can do this because it has no mass – it's all explained by physics.)

Here's another weird thing. Any beam of light approaching closer to a black hole than 2.6 rs will get trapped in the Photon Sphere. This creates the 'shadow' of the black hole. These were rays of light, coming from the other side of the black hole, skimming closer than 2.6 rs, and then getting trapped inside the Photon Sphere. We can never see those rays of light again.

This 'shadow' of the black hole is about 40 billion kilometres across. Viewed from a distance equivalent to the distance between Earth and the M87 galaxy (53 million light years), that shadow is equivalent to the size of a credit card on the Moon.

And the coloured ring? That's the rays of light (from stars on the other side of the black hole) that skimmed just outside 2.6 rs. They then got curved into a ring by the immense gravity of this 6.5-billion-Solar-Mass black hole.

(Actually, it's a lot more complicated than this. For a fuller explanation, check out the Veritasium YouTube video on 'M87, black hole'.)

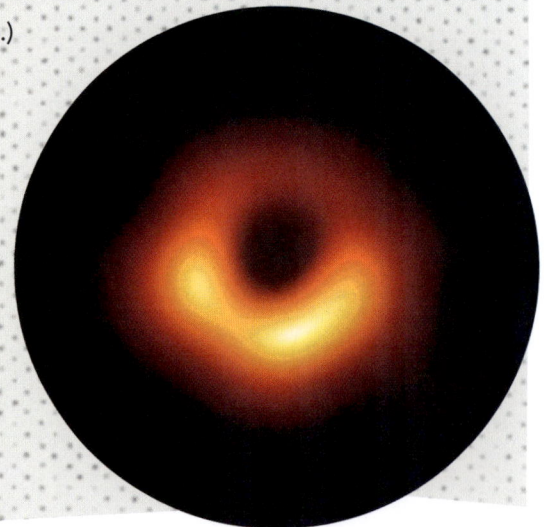

POINT
BREAK

ACCRETION DISC ENERGY

What can happen to stuff when it falls onto an accretion disc, before ending up inside the black hole? This stuff can be moving very quickly – at a significant percentage of the speed of light.

There are two main scenarios: a non-spinning black hole and a spinning black hole.

If matter falls onto the accretion disc of a black hole that is not rotating, about 5.7% of the infalling mass can be converted into energy.

But if the black hole is rotating, that energy can jump to an enormous 42%. (The explanation lies in complicated physics.)

Whether the black hole is rotating or not, there's still a colossal amount of energy available.

It can result in huge jets of matter squirting out of the accretion discs. These 'astrophysical/relativistic jets' have been measured with lengths of over millions of light years, and with speeds of up to 80% of the speed of light.

Sometimes, half of the stuff falling into the accretion disc will get squirted out as an astrophysical jet. But the other half will end up inside the black hole. What happens to it? We don't know. All we know is that it goes into the black hole and never comes out again.

Relativistic jet. The environment around the AGN (Active Galactic Nuclei) where the relativistic plasma is collimated into jets which escape along the pole(s) of the supermassive black hole.

Inner structure of an active galaxy

0.1 light years

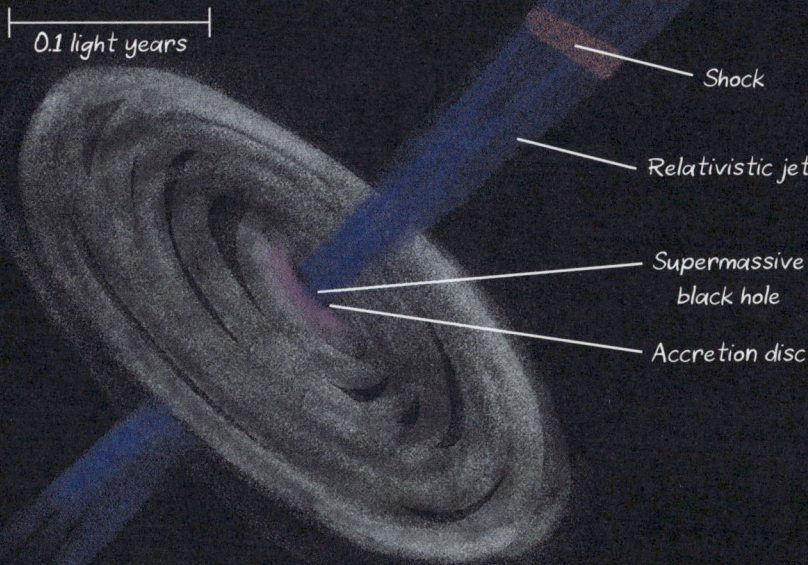

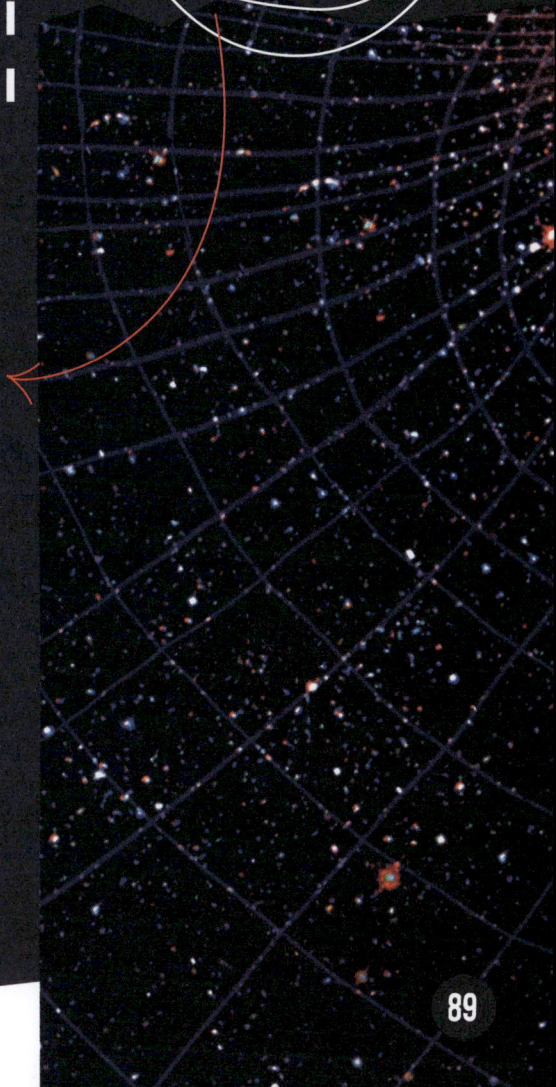

Shock

Relativistic jet

Supermassive black hole

Accretion disc

89

PTEROSAURS, NOT JUST PTERODACTYLS

A cast of Rhamphorhynchus muensteri, a long-tailed pterosaur from the Jurassic period.

Kids love dinosaurs, and pterodactyls are a captivating part of the dinosaur story. Weren't they the trippy version of a dinosaur that had two legs and could fly?

No, actually they did not have two legs. And no, they were not dinosaurs. But yes, they could fly.

And we have worked out how pterosaurs (which include pterodactyls), the heaviest creatures ever to take flight, managed to get off the ground and launch into the air.

PTEROSAUR 101

Pterosaur means 'flying reptile' – or, literally, 'wing lizard'. Pterosaurs existed around the same time as the dinosaurs – from around 220-or-so million to 66 million years ago. But they were *not* dinosaurs. Pterosaurs were a completely different species from dinosaurs.

Pterosaurs were the very first group of vertebrates (animals with backbones) to become fully adapted to powered flight. They dominated the skies of planet Earth for more than 150 million years. Pterodactyls were just *one* of the more than 130 different types of pterosaurs that existed, but for some reason, pterodactyls are the famous ones. Maybe it's because they were quite big, with a wingspan of over a metre, and also were the first pterosaurs to be discovered, in 1784. The word pterodactyl means 'wing finger'. And yes, pterodactyls did fly – just like all the pterosaurs.

Pterosaurs had four legs not just two, and they were neither dinosaurs nor birds, as they descended from completely different ancestors. There were many different types of pterosaur, with many different lifestyles. Like reptiles, they gave birth by laying eggs. They ate insects, fruit, land animals, fish and even other pterosaurs, depending on where they lived and what was available.

Pterosaurs were first described from fossils in 1784, by the Italian scientist Cosimo Alessandro Collini, who discovered the first pterodactyl.

It's always been difficult to find fossils of pterosaurs. Their bones were hollow with a very thin wall, which made them fragile. So, relatively few fossils exist for us to examine today.

Pterosaurs did not have feathers in their wings. Instead, their

Pterosaurs at home

We have gained insights into how some of the pterosaurs lived, thanks to the spectacular discovery of a 120-million-year-old fossil site in northwestern China, which was once a large freshwater lake.

Not only did the scientists find more than 40 pretty good skeletons from a previously unknown species of pterosaur, but they also found five pterosaur eggs that had survived the last 120 million years without being crushed. A later expedition to the same site found several hundred eggs. These are the best preserved pterosaur eggs ever found, and they resemble the soft eggs of some modern lizards and snakes.

It seems that these pterosaurs were gregarious, living in large colonies, and they buried their eggs in moist sand to stop them from drying out.

Comparison of Quetzalcoatlus and Cessna

wings were a flexible membrane of muscle and skin. This membrane stretched all the way from their ankles and hind legs, up along the sides of their rather short body and out to their highly elongated fourth finger. There was also a membrane running between the wrists and the shoulders.

One type of pterosaur, Quetzalcoatlus, was the largest creature ever to fly, with a wingspan of 10 metres and more – bigger than an F-16 fighter. That's three times bigger than today's largest flying animals – the wandering albatross and the Andean condor. On the ground, it stood as tall as a giraffe does today.

There were two major groups of pterosaurs. First, there were the basal, or non-pterodactyloid, pterosaurs. They were smaller, had jaws with teeth, often had long tails, and had an awkward sprawling posture on the ground. Some had wings only 25 cm across. The other group were the pterodactyloids. These were bigger, walked easily on all four limbs, and (as the name implies) included the pterodactyls.

Some of the species had a trunk that made up only one quarter of their body length. The other three quarters was their very long neck and even longer head. If you were their prey, they were the flying jaws of death.

About halfway through the reign of the pterosaurs, about 80 million years after they first appeared, the earliest types of birds emerged – evolving from small two-legged meat-eating dinosaurs. These birds were usually much smaller than pterosaurs.

TAKE-OFF!

Some pterosaurs weighed as much as 250 kg. So the big mystery has always been, 'How did they take off or launch into flight, when they were that heavy?'

Well, scientists have finally worked it out.

Did they climb up a tree and spring out, or head for the nearest cliff and jump off? No. They used millions of years of evolution to lift them off the ground.

First, they had a very strong but very light skeleton.

Second, their membrane wings actually gave more lift than wings with feathers.

And, finally, they had lots of 'haunch/launch' power. They had four limbs – shorter, powerful hind legs and longer front legs that unfolded into wings. This gave them more than double the power of

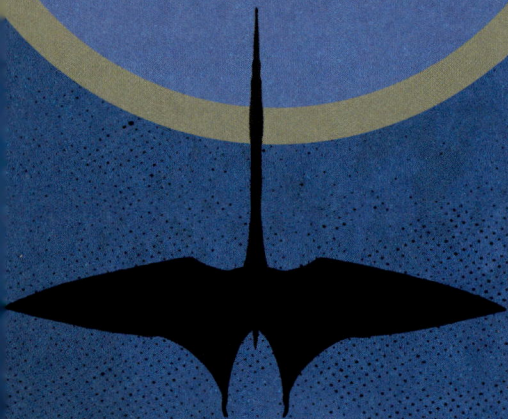

a two-legged animal trying to launch. They would crouch with their rear legs bent, straighten their rear legs to vault upwards, push up with their front legs to add a catapult action, and then finally lift off – all 250 kg of them.

So, in the Great Extinction 66 million years ago, why did all the pterosaurs die out, when the birds did not? I'm afraid nobody knows the answer to that one – yet.

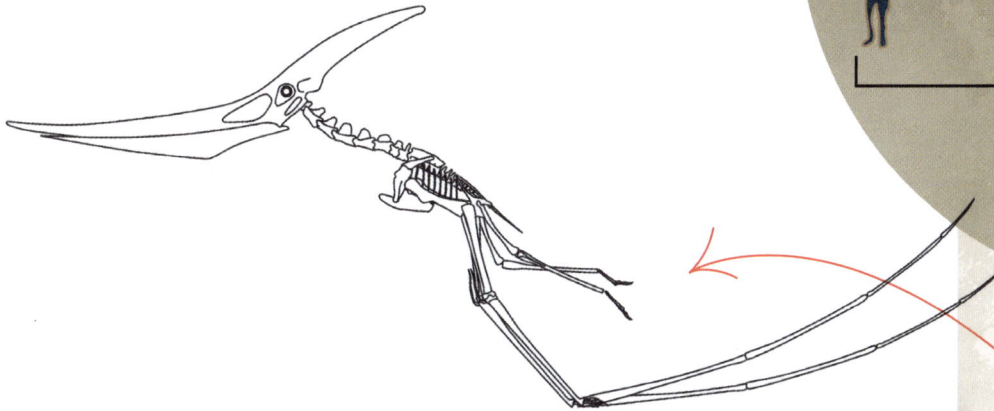

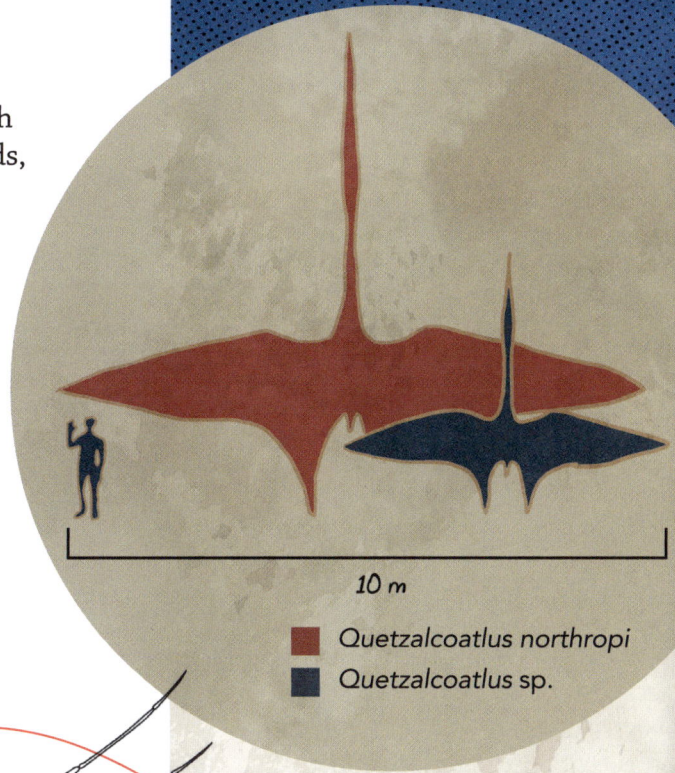

Quetzalcoatlus northropi

Quetzalcoatlus sp.

10 m

Skeleton of a Pteranodon longiceps launching, using all four legs. The first phase of launch is over – the short back legs have done their pushing. This is the second phase of the catapult launch, using the much longer front legs (which are folded up wings now, but will spread open once airborne).

MARATHON RUNNERS' GUT BACTERIA

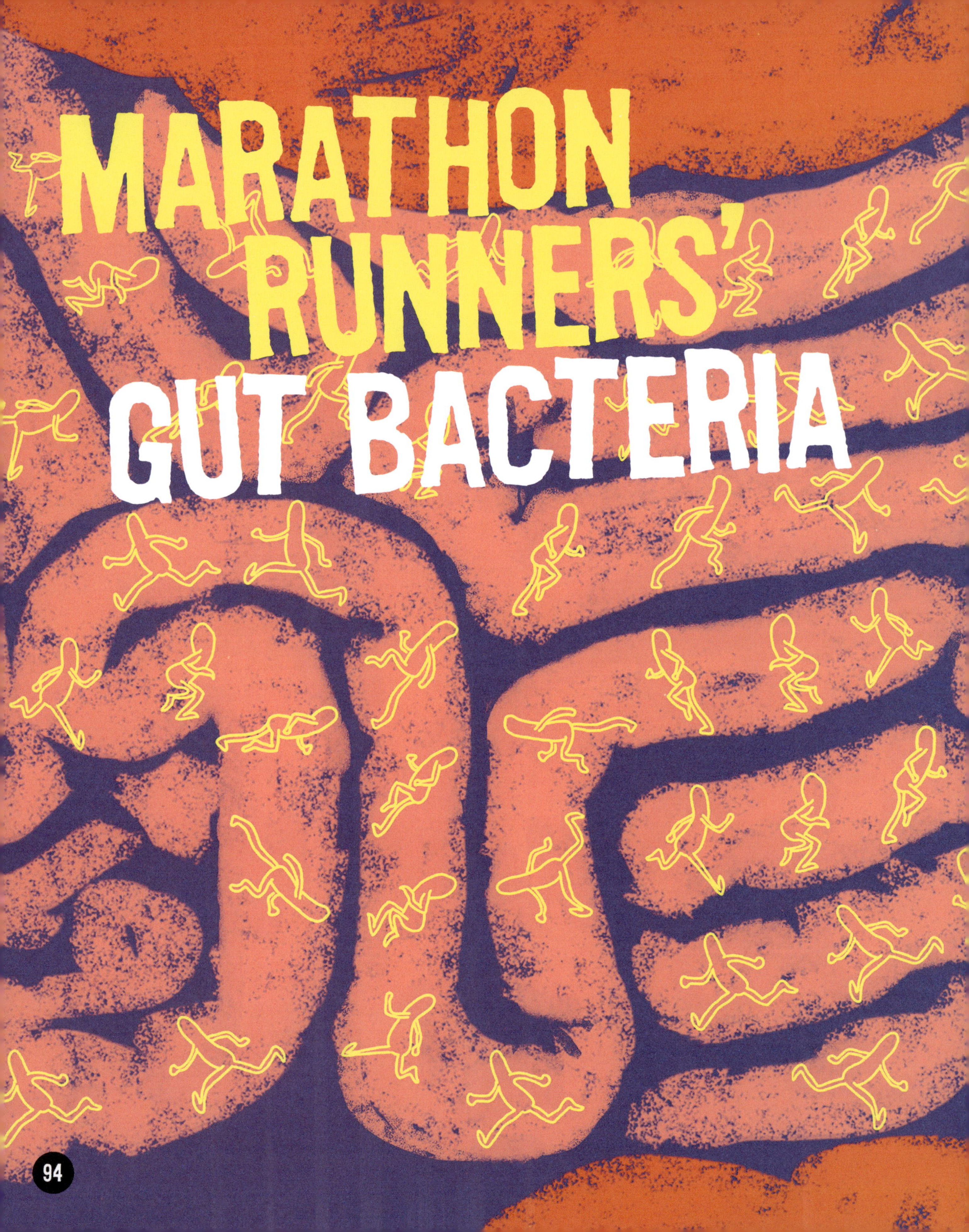

Poo transplants and gut bacteria squirted into the public consciousness about a decade ago. We were beginning to truly appreciate just how influential the bacteria that live on us, and inside us, are. Our human body would be so unhealthy without our bacterial friends – all 200 grams and 38 trillion of them. We realised that bacteria and humans are in it together, for the long haul.

And speaking of long distance, some really surprising recent research found that the bacteria in your gut can help you win a marathon! The bacteria can literally turn an unwanted byproduct of exercise (lactic acid) into energy (adenosine triphosphate, or ATP).

MICROBIOME 101

Let's start explaining how bacteria can help you win a marathon by looking at the concept of 'cells'.

Every one of us started off when an egg cell was fertilised by a sperm cell. That single fertilised egg cell then grew and split many times over the next nine months, before turning into a baby. That first cell turned into liver cells, brain cells, lung cells, kidney cells and many more quite different cells – eventually amounting to 30 trillion of them!

But the bacterial cells your body also carries can beat that number – about 38 trillion bacteria! That's right. As you were being delivered from your mother's uterus, bacteria came on board and bred mightily, using you as their home. They live on your skin, inside your gut, and in a few other places. These bacteria are a lot smaller than your human cells, weighing only 200 grams or so in total. The vast majority live inside your gut.

Don't cringe when you think about being loaded with more bacterial than human cells! We have all been subjected to decades of misleading propaganda telling us that all bacteria are evil, and that we should kill 99.9% of them. Wrong. We need bacteria, and they need us.

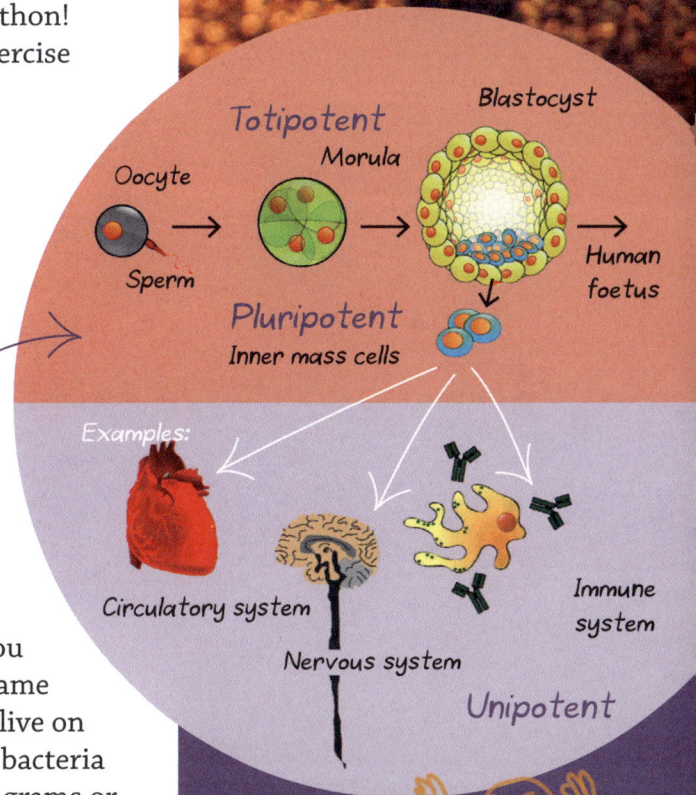

Oocyte

Sperm

Totipotent

Morula

Blastocyst

Human foetus

Pluripotent

Inner mass cells

Examples:

Circulatory system

Nervous system

Immune system

Unipotent

Our bodies are made up of more than just cells

An adult has about 30 trillion self-generated cells. Surprisingly, about 84% of your cells are red blood cells. However, red blood cells are some of the smallest cells in your body, so they make up only about 4% of your total cell weight. At the other extreme, muscle and fat cells are very big, so while they make up just 0.2% of your total cell count, they make up about 75% of your total cell mass.

But most of 'you' is not cells. The water that makes up about 60% of your body weight is not cells, and neither are hair, fingernails, collagen, elastin and so on.

These bacteria that live on and inside you are usually your friends. In most cases, they are essential for your good health. Between them, these 500–1,000 different species of bacteria in your gut carry about 100 times as many genes as your own personal DNA. You can think of these bacteria as an extra metabolic organ that has two properties. First, this 'organ' is exquisitely attuned to your own special and individual physiology (which is different from everybody else's). Second, this 'organ' can do things that your own physiology cannot.

For starters, think about plant polysaccharides. Your personal gut physiology cannot process some of them, so (initially, at least) those specific plant polysaccharides are indigestible. But the microbiota (the collection of all the bacteria in your gut) *is* able to break down those plant polysaccharides – and provide nutrition to both you and them. Everybody benefits!

If you had absolutely no bacteria in your gut, you would probably need to eat twice as much food and weigh about two-thirds as much. You would be a thin, sickly version of yourself, with a much-impaired immune system. Your bacteria, and their byproducts, are essential to your good health and wellbeing.

MARATHON RUNNERS' MICROBIOME 101

So that's the background to our cells and our 'microbiome'. Now, let's race on to our marathon runners – how can gut bacteria help them run faster?

It's a two-part answer – and yes, it came from a study that involved poo testing.

First, long-distance runners have slightly different gut bacteria from 'regular' people (people like me, who have never run a marathon). Second, these novel bacteria seem to give them an extra energy boost.

Think about 'energy' as being like 'money'. If you want to do anything in our 21st-century world, you almost always have to spend money – our 'fiscal currency'. In a similar way, if you want to do anything in your body, you have to break down an energy molecule. This energy molecule is the 'energy currency' of your body. It's ATP.

MY FAVOURITE MOLECULE

The energy molecule you have to 'spend' in your body to do anything is called ATP, which stands for adenosine triphosphate.

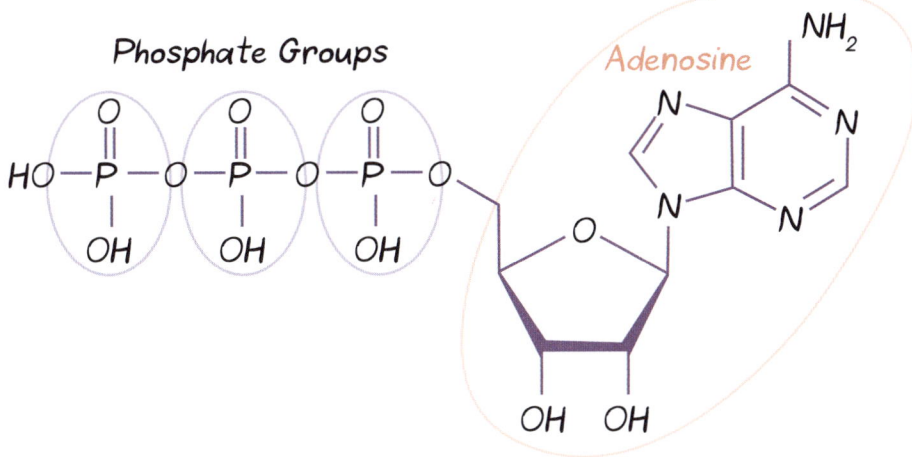

Phosphate Groups   **Adenosine**

Whenever you blink an eyelid, or make urine with your kidneys, or move your leg muscles, you burn up some ATP (and turn it into ADP, adenosine diphosphate).

Adenosine triphosphate is made up of an adenosine part and three phosphates (hence its name). There is a lot of energy embedded in how each of the three phosphates is stuck on. The first phosphate is stuck onto the adenosine, the second phosphate is stuck onto the first phosphate, and the third phosphate is stuck onto the second phosphate. The phosphates can be removed only in sequence – the third (or outermost), followed by the second (or middle), and finally the first (or innermost). The act of removing a phosphate releases some energy, which you can then use to contract a muscle, pump potassium into a cell, etc.

Now, here's the okay news. For the two phosphates that are stuck on closest to the adenosine, you have to put in a moderate amount of energy to get back somewhat more energy. Actually, that's quite good news – you are getting back more energy than you put in.

And there's even some extra good news. That last phosphate, which is stuck on right at the end of the phosphate chain, gives you a much better deal. All you have to put in is a tiny amount of energy, and in return you get back much, much more energy!

In the vast majority of cases, your body gets its energy by just playing with that third phosphate. After all, that makes sense. You get the most bang for your buck by shifting that end phosphate – put in a small amount of energy, and get a whole lot more energy back.

Elite athletes' poo

Dr Jonathan Scheiman and his team examined the poo from 15 runners, mostly every day in the week before and then the week after the 2015 Boston Marathon. They found an increase in one genus of bacteria, *Veillonella*, in their poo samples (as compared with non-runners). The levels of this bacterium spiked high after an intense training session, and even higher after the actual marathon.

He separately found that mice he had fed orally with tidbits of *Veillonella* were able to run 13% further on a treadmill than mice that had not eaten the extra *Veillonella*. (Any athlete will tell you that a 13% improvement is enormous. And I'll bet they'd be prepared to spend a lot for an energy boost of that size. So I'm guessing that this bacterium could be sold in the future as a 'performance-enhancing' product – perhaps a probiotic?)

A different Irish study found a spike in *Veillonella* levels in rugby players after exercise.

So it seems elite athletes truly are different from the rest of us, inside and out – even down to their poo!

Your own weight of ATP!

At any given instant, you are carrying between 100 and 250 grams of ATP molecules. These molecules are scattered across your body, wherever work is being done.

ATP is continually being broken down to ADP to provide energy, and is almost immediately recycled back up to ATP again. This recycling happens in each and every single second that you are alive – even when you're asleep. Over an average day, you will break down, and then rebuild, roughly your own body weight of ATP.

P_i
+
ADP
+
Huge amount of energy
Tiny amount of energy
+
ATP

Taking off the end phosphate means you turn adenosine triphosphate (with three phosphates) into the less energetic version – adenosine diphosphate (with just two phosphates). And then, inside the mitochondria which are inside your cells, the low-energy ADP is recycled back into the high-energy version, ATP. This is one of those very lovely 'circles of life', and it's why ATP is my Favourite Molecule.

(By the way, there are unusual situations where your body will go to the trouble of ripping off the second phosphate, leaving you with AMP, or adenosine monophosphate, but it's not very common.)

YOU NEED O_2 TO MAKE ATP

And where does the external energy come from to rebuild your ATP?

Mostly from the sugars in the food that you eat *and* from the oxygen that your blood delivers to help break down the sugars, releasing their potential energy. This is called the glycolosis pathway.

If you're fit and well, you probably have a robust blood supply that can deliver lots of oxygen from the lungs to the muscles. For marathon runners, it's the leg muscles that count. With a good oxygen supply, your muscle cells will use 1 molecule of glucose ($C_6H_{12}O_6$) and 6 molecules of oxygen to change 38 molecules of low-energy ADP into 38 molecules of high-energy ATP. (For complicated reasons, it's usually a bit less than 38, but we can ignore that for now.) As part of this process, you also make 6 molecules of carbon dioxide, which you breathe out.

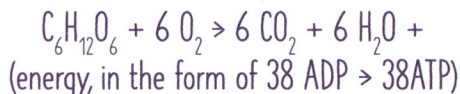

$$C_6H_{12}O_6 + 6\ O_2 \rightarrow 6\ CO_2 + 6\ H_2O +$$
(energy, in the form of 38 ADP → 38ATP)

But if you haven't done enough training, your lungs and blood vessels can't deliver quite enough oxygen. In this case, 1 molecule of glucose gets turned into just 2 lousy molecules of ATP – not 38. And, even worse, instead of making 6 molecules of carbon dioxide, which you breathe out, you make 2 molecules of lactic acid.

$$C_6H_{12}O_6 \rightarrow 2\ CH_3CH(OH)COOH + \text{(energy, in the form of 2 ADP → 2ATP)}$$

Lactic acid is what generates the unpleasant burning sensation of 'lactic acidosis' in your muscles, and can cause shaky legs after very intense workouts. For these reasons, having lactic acid in

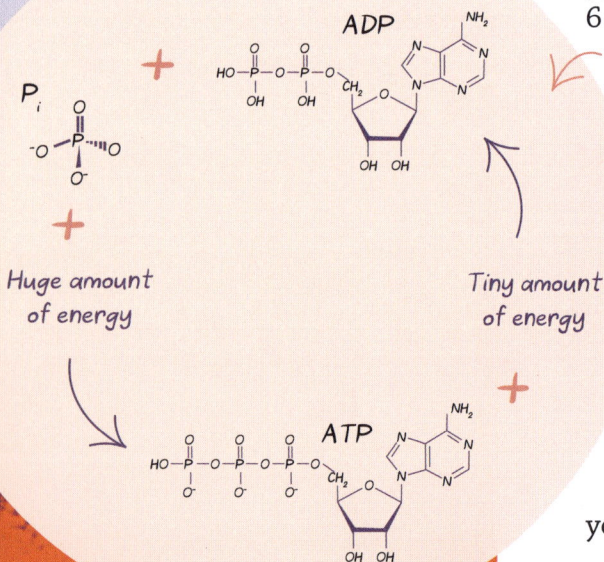

the bloodstream is, for most of us, always a disadvantage. But in elite marathon champions, it can have an upside (because of their abundance of *Veillonella* gut bacteria).

LACTIC ACID GIVES 'FREE ENERGY'

Most people vaguely know that lactic acid can affect your muscles.

But what I didn't know was that lactic acid might be able to leave the bloodstream and do a little tour of duty of the gut, recruiting its bacterial friends, *Veillonella*, to help it assist marathon runners.

Remember that, compared with average sedentary humans, marathon runners have lots more of a special type of bacteria, called *Veillonella*, in their gut. When lactic acid enters the gut, these bacteria break it down, turning it into a short-chain fatty acid – propionate. This fatty acid then leaves the gut at the far end – the sigmoid colon and the rectum – to enter the bloodstream, bypass the liver, and end up in the muscles, where it's turned into a little extra energy.

So, the astonishing result is that these bacteria, which are more plentiful in athletes, actually give extra energy to these athletes. They convert an unwanted byproduct of exercise into the molecule of energy, ATP.

This gives athletes yet another potential performance enhancer – with a significant yuck factor! Will 'faecal doping' (loading the rectum with fatty acids before a big race) be the Next Big Fixer? However, before people start doing this, they should realise that the metabolic effects of the chemical, propionate, are actually quite complex (insulin resistance, etc).

CHICKEN OR EGG

Well, we're not exactly sure yet which came first.

Did some people accidentally have *Veillonella* bacteria in their gut, which gave them an extra advantage when it came to running a marathon? Or was it the hard training that flooded the gut with lactic acid, which the *Veillonella* bacteria came to eat?

One thing is for sure – we're just beginning to understand all the things those 38 trillion bacteria living in and on us do, especially the ones in our gut.

Who'd have thought there was so much more to poo than a bunch of bum jokes?

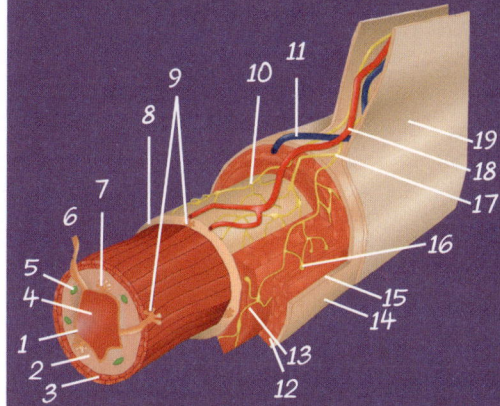

General structure of the gut wall
1. *Mucosa*: Epithelium
2. *Mucosa*: Lamina propria
3. *Mucosa*: Muscularis mucosae
4. Lumen 5: Lymphatic tissue
6. Duct of gland outside tract
7. Gland in mucosa
8. *Submucosa* 9. Glands in submucosa 10. Meissner's submucosal plexus 11. Vein
12. *Muscularis*: Circular muscle
13. *Muscularis*: Longitudinal muscle 14. *Serosa*: Areolar connective tissue 15. *Serosa*: Epithelium 16. Auerbach's myenteric plexus 17. Nerve
18. Artery 19. Mesentery

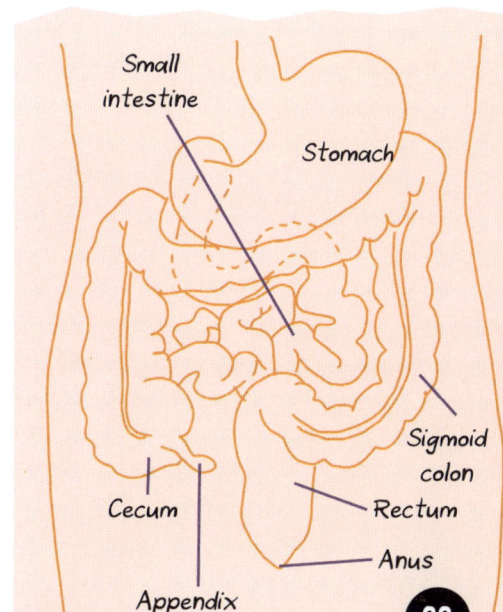

Small intestine

Stomach

Cecum

Sigmoid colon

Rectum

Anus

Appendix

COMBUSTION ENGINES –

THE BURNERS GO BUST?

I've been a motorhead, and loved cars, even before I could drive. As a student, I spent hours hooning around on the winding country roads behind Wollongong, instead of going to my university lectures. We called driving just for the sheer pleasure of it 'going for a burn'. I have test-driven 4WDs through most of the deserts of Australia, and have spent about two years in total driving through the Outback.

So I *love* driving – but I'm not blinded by the exhaust smoke to its environmental impact.

When I wrote my first story on Climate Change, it was so long ago (1981) that it had a different name – the Greenhouse Effect.

INTERNAL COMBUSTION ENGINE ABANDONED

So that's why I was thrilled (from an environmental point of view) to read the headline: 'Daimler abandons Internal Combustion Engine development to focus on EVs'. (EVs are Electric Vehicles.) Daimler makes Mercedes-Benz. Other European car manufacturers are doing the same.

For me, the key word in that headline is 'development'.

The internal combustion engine went through an incredible amount of development in the 1980s and '90s, thanks to a mixture of environmental concerns, and cheap and powerful computer technology. You only have to look at the difference between a small 1970 car (when I was a single bloke) and a large 1995 station wagon (family man) to see the huge leap.

The 1995 vehicle was a vast improvement. The newer engine had about twice the power. The new car body was twice the weight and provided much better safety – both active

My old 4WD

End of the burners

One motoring journal summed up the transition from internal combustion to electric engine in the headline, 'VW: 2040 last burner off the assembly line'.

For me, the word in that headline that stood out was 'burner'.

Fossil fuels are converted into energy by burning. In Australia, car emissions standards are high, and we have the advantage of a small population in a big country. Even so, in the cities of Sydney and Melbourne, the estimated number of deaths from fine-particle pollution in 2015 was over 2,200.

Bertha Benz, automotive pioneer

and passive. Active safety is stuff you have some degree of control over, such as steering (which can pull you back onto your side of the road), the accelerator (which backs off when you lose wheel grip), the brakes (anti-lock braking), etc. Passive safety is stuff that is built into the car, and over which you have no control – such as intrusion bars in the doors (to stop another car coming into your passenger compartment through your door), air bags, Crumple Zone Technology (so the body collapses, but you live), low centre of gravity (for better handling), etc.

Now, you'd think that with all its extra power and weight, the 1995 vehicle would be a total fuel guzzler – using four times as much (twice the weight, twice the power). But no – its fuel economy was virtually identical, while its engine emissions were about 80% lower!

This is what 'development' means – continuous improvement on all fronts.

But the major European car makers are stopping development work on internal combustion engines. That covers petrol, diesel and gas. They'll still make these engines for a while – but their Research & Development (R&D) budget is shifting across to electric cars. Daimler's current plan is that the latest generation of internal combustion engines (e.g. inline six-cylinder engines for the E- and S-class, etc) will be their last generation. Their main focus is now on 'electrification, electric drives and battery development'. Daimler said, 'The car will change more in the next ten years than in the 100 years before.' This is an incredible goal!

So it's not just Daimler that is switching its vast resources from the internal combustion engine to the electric vehicle. Volvo and Volkswagen have also announced that they are putting their future R&D into electric vehicles.

CAR ENGINE – BIRTH TO DEATH

Back in 1888, automotive pioneer Bertha Benz (business partner and wife of automobile inventor Karl Benz – as in Mercedes-Benz) took the world's first ever long-distance road trip in a car – 100 km to visit her mother. It took her a full day. Like many road trips, it had a few ups and downs. There were no roads, no petrol stations and no roadside assistance.

INTERNAL VS EXTERNAL COMBUSTION?

Every now and then, you might hear a car engine described as an 'internal combustion engine'. So how is that different from an 'external combustion engine'? The first word in the name gives you a big hint.

What's common to both types of combustion engines is that they have a bunch of cylinders (usually more than one). Inside the cylinders are pistons that move up and down. That up-and-down motion is converted into the rotary motion of the crankshaft, which makes the axles spin, and then the wheels turn.

And what makes the pistons travel up and down?

It's hot gas, under pressure. The peak pressure inside a cylinder can reach between 20 atmospheres (engine under light load) to 100 atmospheres (racing engine under heavy load). By comparison, the pressure in car tyres is usually about 2 atmospheres. Because the gas is loaded with the energy of pressure, it can push the piston.

In both the 'internal' and 'external' types of engines, the heat from burning is used to turn a liquid into a gas. That's the 'combustion' part, which is common to both. When a liquid turns into a gas, the volume increases enormously. For water, the increase is about 1,700 times. The liquid can be petrol or diesel (internal combustion) or water (external combustion steam engine).

So the key difference between 'external' and 'internal' combustion is where the heat is generated. In an external combustion engine, the combustion happens outside the cylinder, while in an internal combustion engine, the combustion happens inside the cylinder.

A steam-train engine is your classic case of an external combustion engine. A fire (wood or coal) is lit under a metal container holding water. After a while, enough heat has been added so that the water (a liquid) turns into steam (a gas). The steam, which is under pressure, is fed into a cylinder, and then pushes a piston. The piston then turns the steel wheels of the train. I am always amazed that just one person shovelling coal can provide enough energy to push a 500-tonne train along steel rails at 100 kph.

In an internal combustion car engine, a mist of very fine droplets of petrol is squirted into the cylinder. The spark plug ignites the petrol-air mixture, generating gas at pressure, which then pushes the piston down.

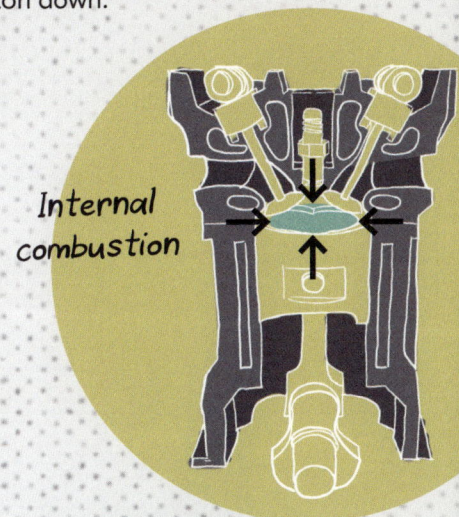

Internal combustion

External combustion

103

Despite these obstacles, as well as multiple breakdowns, Bertha made it. She was the midwife for the birth of the motor car, in 1888. The descendants of Bertha's burner have been running for over a century. But now the internal combustion engine, in motor vehicles, is on its deathbed – according to the 2019 news from Europe.

I read about the demise of the fossil fuel–burning car in the well-respected German motor magazine *Auto Motor und Sport*. Germans love their cars. They invented the motor car, their drivers go through a far more rigorous driving test than ours, and some of their freeways, or autobahns, have no speed limit. This magazine has the motto 'Petrol in their blood'. Every fortnight, it gives a high-quality, up-to-date auto-market summary that includes critical analysis, observation of the latest trends, service tips and environmental topics. So if something's afoot in the car industry, it'll be in this mag.

Recently, some 20,000 of its readers voted on the question, 'Should automakers really stop developing combustion engines?' Now, remember that this is a sample of total petrolheads. As you might expect, some 29% voted 'No'. But, excitingly, an even larger percentage, 47%, voted 'Yes. The time of the burners is over. End of story.'

THE FUTURE OF THE CAR

So there you have it. Big it up for the fully electric car that is coming your way. It comes with advantages that well and truly make up for any anxiety about recharging.

One big one is that there will be fewer deaths from air pollution, and of course there'll be lower greenhouse-gas emissions.

And electric vehicles can also play a dual role – being part of the electrical grid. There is the nice coincidence that 20 kilowatt hours (kWh) will either run your car for 100 km, or run your house for a day. So, your future car will not only provide transport, but will spend the rest of its time plugged into the grid – either storing or delivering electricity.

I'll always have a soft spot for my old burners, but they'll live on in my memories and photo albums and the pages of history, where they belong.

POINT BREAK

ELECTRIC DREAMS

USA's electric trucks – big

The USA has about 15.5 million trucks. About 2 million are semitrailers/semis (or tractor-trailers/18-wheelers, or articulated lorries/artics, depending on which country you're in). About one-eighth (250,000) get replaced each year. Medium and heavy trucks create about 8% of the USA's total greenhouse-gas emissions.

The average diesel semi costs about US$1.38/mile to operate. But leading electric-car manufacturer Tesla estimates that its electric semi will cost US$1.26/mile. Though the upfront cost will be higher.

The roll-out of electric trucks is already starting (Daimler, Kenworth, Volvo, Tesla, etc). It will happen in three phases – medium-duty trucks and vans, then big semis for regional hauling, and finally big semis for long-haul.

USA's electric trucks – small

In the USA, the top seven bestselling cars are actually light trucks and SUVs (sports utility vehicles). Light-duty vehicles like these account for over 60% of transportation greenhouse-gas emissions. In the last decade, these vehicles have increased the USA's emissions more than was done by heavy industry, planes or cargo ships. In fact, if SUV drivers around the world were a single nation, they would rank seventh in the world for carbon emissions.

One new electric car company, Rivian, has a contract with Amazon to deliver 10,000 electric delivery vans by 2022, and another 90,000 by 2030. The company, realising that only 10% of US utes/pickup trucks are used for work, is also planning to release its first 'glamorous' electric pickup truck by the end of 2021.

In 2019, the bestselling vehicle in the USA was the Ford F-150 – 896,526 sold. Ford is planning to release its first electric F-150 in 2021.

China's electric vehicles

The production of electric vehicles in China is ramping up very quickly, due to stringent new-energy vehicle production quotas being rolled out. China's long-term goal is to become one of the major producers of electric cars in the world. In 2018, the Chinese company BYD made and sold more electric cars than any other manufacturer in the world – 248,000. China buys more electric cars than any other country – 1.25 million in 2018. By the end of 2018, China had about 342,000 public charging points for electric cars, while the USA had 67,000.

The megacity of Shenzhen (population of 12 million) has the world's first and largest electric bus fleet – 16,000 buses. A two-hour charge will power a bus for its typical daily run of 200 km. Essentially all of Shenzhen's 22,000 taxis are electric.

Part of the reason for China's move to electric vehicles is to do with the nation's energy security – moving away from oil (70% of which is imported) to electricity (which can be made onshore).

Norway's electric target

Norway has a target of selling only emission-free cars by 2025.

In April 2020, the total number of new cars sold dropped by 34% (as compared to April 2019) due to COVID-19. However, while the sales of all petrol and diesel cars dropped by 60%, electric-car sales dropped only slightly, and made up 50% of all car sales.

Japan's carbon-free goal

Subaru announced in January 2020 that all its vehicles by the mid-2030s would be electric. This is part of Japan's plan to create a carbon-free society.

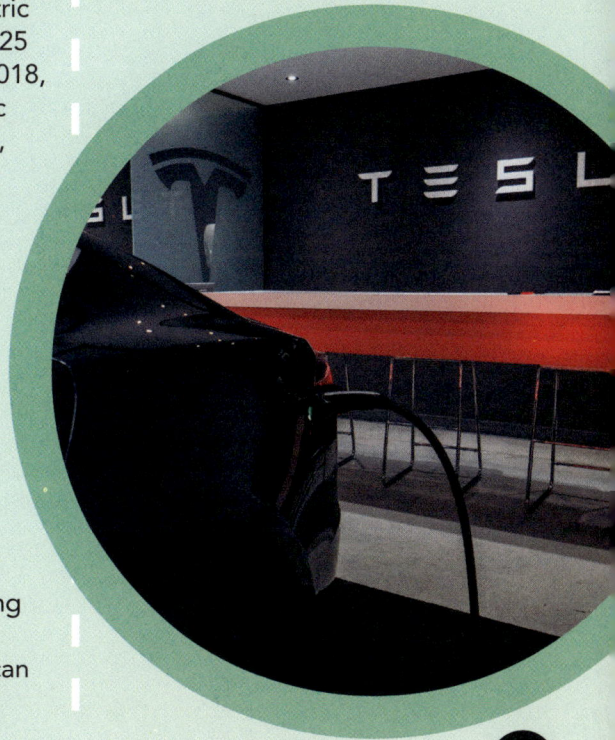

TIME IS RUNNING OUT FOR SAND

You might say that sand is as common as mud – after all, beaches are full of the stuff. Romantically speaking, there are roughly as many grains of sand on planet Earth as there are stars in the entire known Universe (which is pretty poetic). So how can it possibly be true that we are running out of sand!

Sadly, it is true. In fact, Dubai, a desert city, has already run out of useful sand. They are importing sand from Australia to keep their construction industry humming along.

We're not short of desert sand. But for building purposes, those grains of desert sand are useless – they are too smooth to grip to each other. For construction (the biggest user of sand on the planet) you need sand with a rougher surface – usually from rivers and oceans.

SAND & TIME

We've been using sand for 5,500 years. Back in 3,500 BCE, the Mesopotamians and the Egyptians could turn sand into glass.

Since then, sand has become a key ingredient of, and absolutely essential to, our modern society. Between them, sand and gravel are the most extracted materials on Earth, beating both fossil fuels and iron ore. We use more of them than any other resource – apart from water and air.

Sand is used in agriculture (to help soils drain water away more easily), in the manufacturing industry (to make abrasives such as grinding wheels and sandpaper, and moulds for casting hot liquid metals, and so on), and in hydraulic fracturing or fracking (where sand grains keep fissures open in underground coal seams, to let natural gas out, so we can burn it and create both heat and more greenhouse gases). Sand is also used to make window glass and optic fibres.

But the number-one user of sand is the construction industry.

THE AVOGADRO NUMBER, STARS AND SAND

Hang onto your hats, here comes a nice coincidence!

In 1811, the Italian scientist Lorenzo Romano Amedeo Carlo Avogadro came up with his insights into the concept of 'molecules'.

First, as you probably guessed, with so many names he was minor nobility – he was the Count of Quaregna and Cerreto.

Second, his thinking was pretty advanced for the time. In the early 1800s, science had not even proved the existence of atoms, much less groups of atoms, which today we call molecules.

His most important insight, now known as Avogadro's Law, was: 'Equal volumes of gases under the same conditions of temperature and pressure will contain an equal number of molecules.'

He also has his own number named after him. It's the Avogadro Number, and it's $6.02214076 \times 10^{23}$. It's a very big number (after all, there are 23 zeros in it).

It's the number of particles/molecules in 18 grams of water (H_2O), or 2 grams of hydrogen gas (H_2) or 32 grams of oxygen gas (O_2). Chemists use the Avogadro Number all the time.

And here comes the coincidence: there is roughly an Avogadro Number of grains of sand on Earth, and of the stars in the entire Universe.

Why? Dunno. Maybe you'll win a Nobel Prize if you find out why …

| 18 gm of H_2O | = | 32 gm of O_2 | = | 2 gm of H_2 | = | Grains of sand | = | Stars in the Universe |

SAND 101

But first, what actually is sand? Unfortunately, the definition is kind of fuzzy.

Sand is part of what the construction industry calls 'aggregate'. And, by the way, the definition of the dimensions of sand varies from country to country.

If the particles that make up 'aggregate' are really small (less than 0.06 mm in diameter), the product is called 'silt'. If the particles are a bit bigger (between 0.06 mm and 4.7 mm), that's your classic 'sand'. And if the particles are bigger than 4.7 mm across, they're called 'gravel'.

All around the world, sand is continually being generated by rocks rubbing up against one another – usually in rivers or oceans. Amazingly, before these particles arrive as sand at your local beach or desert, they go through as many as seven generations of refinement. They get rubbed against each other, then buried, then rubbed some more and buried – a process that is repeated up to five more times.

There are many different types of sand.

The sand at your local beach is usually based on crystalline silica – quartz, or silicon dioxide, SiO_2 (that's one atom of silicon joined to two atoms of oxygen). This is the most common type of sand.

The second most common type of sand is based on aragonite, or calcium carbonate, $CaCO_3$. This type of sand originally started out as the shells of long-dead sea creatures, such as coral and shellfish.

Some of the black sand of the world carries magnetite, or iron oxide, Fe_3O_4. And the white sands of New Mexico in the USA are based on a different chemical, gypsum, or calcium sulphate, $CaSO_4$.

So the different sands can come in many different colours, depending on the chemicals inside. These different chemicals can give sands many other different properties – abrasiveness, the ability to take and hold a shape when wet, filtering ability, a rough surface that's ideal for use in construction, and so on.

And there's another ingredient that differentiates sands: the physical process that actually turns rock into sand. In other words, good old erosion.

Desert sand is very smooth and rounded. This is because the grains of sand have been blown around by the wind for millennia, which has worn off all the rough edges. This rubbing together of the sand grains happens at relatively high speed – partly because sand grains aren't heavy, and wind is fast. This is what makes desert sand too smooth for making concrete.

The construction industry needs grains of sand that are less spherical and more bumpy. This bumpiness helps them lock into other rough grains of sand. Marine sand – from oceans and rivers – fits the bill.

The bottom line is that there are many different types of sand, but only a few are suitable for the construction industry.

CONSTRUCTION SAND

The construction industry is the hungriest user of sand. The average American home uses 100 tonnes of sand, gravel and crushed stone.

Polishing

Sand is a good example of how different mechanical processes give different results. When I was being taught how to wax a car by hand, I was shown how the car paint looked very different if I used high-speed rubbing with low pressure (better), or low-speed rubbing with high pressure (worse).

So smooth desert sand is what you get with high-speed rubbing and low pressure (from the wind), while rough marine sand comes from low-speed rubbing and high pressure (from the moving water). Another reason marine sand is rougher is because water tends to stick to each grain of sand, providing a buffer that lessens the impact when sand grains collide.

Burj Khalifa

If you include the road in front of the house, that adds another 100 tonnes. One kilometre of a single lane of highway uses 25,000 tonnes of sand. Four lanes (two in each direction) use 100,000 tonnes. In 2013, China built 146,000 km of road – you can see the need for sand is growing ever greater.

In 2012 alone, the world made enough concrete to build a wall some 27 metres high, 27 metres wide, and long enough to loop around the planet at the Equator.

China is on an infrastructure binge, and so, between 2011 to 2013, it used more sand than the USA did over the entirety of the 20th century. In 2016, China accounted for one fifth of the world's sand imports.

BRINGING SAND TO THE DESERT

When Dubai constructed the Palm Jumeirah, an artificially built set of sand islands shaped like a palm tree, it used some 385 million tonnes of its own marine sand, from the ocean.

After the Palm Jumeirah came another two sets of artificial islands. One of these sets of islands, The World, represents a map of the world. It alone used up 450 million tonnes of sand.

Unfortunately, these three sets of artificial islands used up all the marine sand around Dubai. According to a United Nations report, the dredging also caused massive marine environmental damage.

So when Dubai built the world's tallest building, the Burj Khalifa, some 828 metres high, the city imported all the sand it needed (for the high-strength concrete) from Australia.

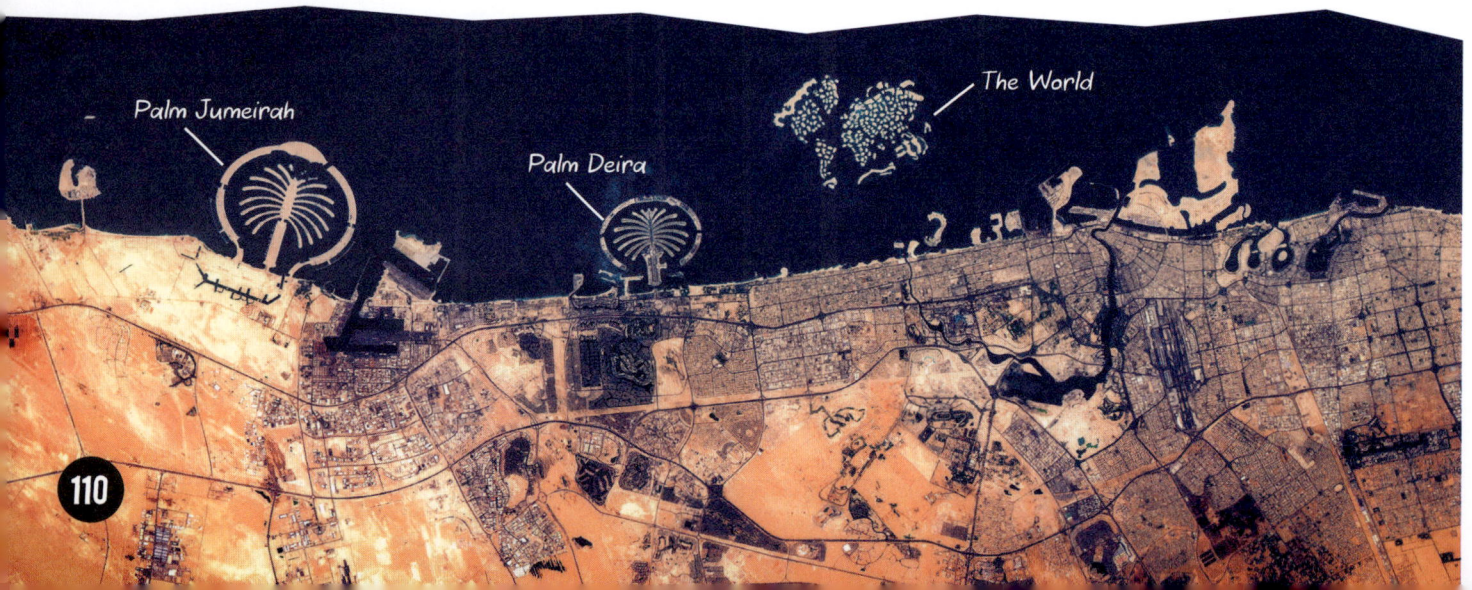

Palm Jumeirah

Palm Deira

The World

SAND MINING

So, how much sand do we mine each year? Surprisingly, we do not have a good answer.

First, we usually don't measure how much we mine. One overview of 443 scientific papers dealing with the mining of sand found that only 38 of them (that's fewer than 10%) measured how much sand was being mined.

Second, when we do measure sand mining, we don't do it well. There are huge gaps in the documentation.

Just look at Singapore to see how hard it is to get accurate estimates of sand use. Since 1960, the population of the city state of Singapore has grown from about one and a half million to five million inhabitants. The land area has increased by about 20% – some 130 km^2. This extra land came mostly from dumping aggregate at the seaside, to reclaim land from the ocean.

In the decade between the years 2006 and 2016, Singapore reported that it had imported some 80 million tonnes of sand from Cambodia. Yet Cambodia reported that it had exported less than 3 million tonnes of sand to Singapore. Nobody can properly explain the missing 77 million tonnes – 96% is an unbelievable amount to 'lose' as a rounding error!

The third reason for our lack of knowledge is blatant theft. Who woulda thunk that sand is so desirable!?

Illegal sand mining is out of control in some 70 countries worldwide. This ranges from a few people in a boat scraping up the local underwater sand with a bucket, to illegal dams and the use of

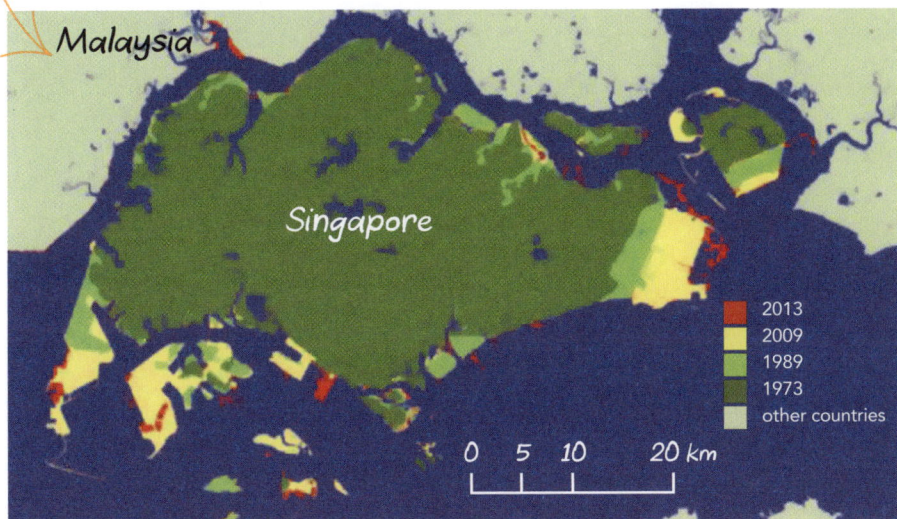

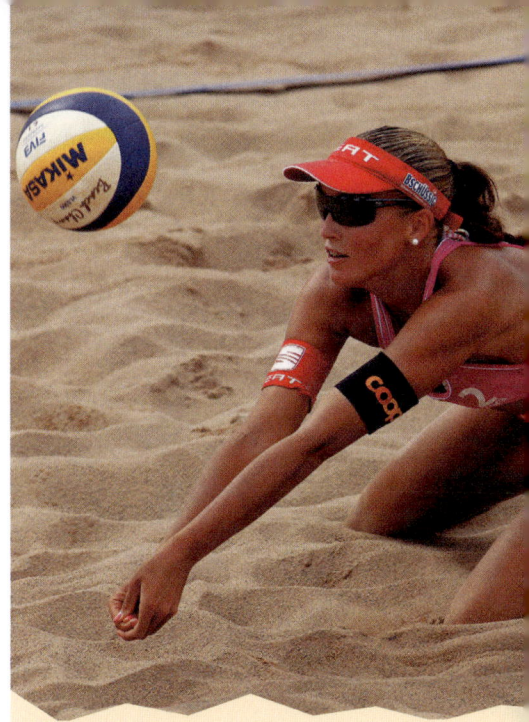

Malaysia

Singapore

2013
2009
1989
1973
other countries

0 5 10   20 km

Sand for sport?

Shifting from work to play, what about sand as used in sport?

For beach volleyball, you need very different characteristics for your sand from that used in construction.

Regular beach sand is too hard for volleyball players, causing them a whole range of impact injuries – from broken fingers to torn hamstrings. After complaints by Canadian players at the 1996 Olympic Games in Atlanta, new specifications were set, stipulating that beach volleyball sand must have a 'soft' surface that gives way easily to pressure.

For the 2016 Beach Volleyball World Tour Final held in Toronto, some 1,360 tonnes of the right sort of sand were imported from Hutcheson's quarry two-and-a-half-hours' travel to the north. It required about 35 semitrailers.

And (of course) you need quite different types of sand again for golf bunkers and for horse dressage.

What's the difference between river and ocean sand?

The big advantage of sand from freshwater rivers is that it doesn't have any significant amounts of salt in it.

Sand from the ocean has to be washed to remove the salt – otherwise, a building made from ocean sand might later collapse from corrosion.

STOP SAND MINING

heavy construction equipment such as bulldozers and big trucks to extract sand unlawfully.

And it's serious business. In India and Kenya alone, in the last decade hundreds of people have been killed in battles between residents and outsiders who were stealing the local sand.

This is why we just don't have accurate numbers for sand mining. But overall, a reasonable guesstimate is that we mine about 50 billion tonnes of aggregate each year for the construction industry. Conservatively speaking, about 10 billion of that aggregate is sand. (By the way, 50 billion tonnes is roughly twice the weight of sediment carried by all the rivers of the world, each year.)

EFFECTS OF SAND MINING

As you would imagine, when you remove massive amounts of sand from its native environment, you have massive flow-on effects. We've removed so much sand from rivers that you can easily see these effects with space satellites.

The Vietnamese government reckons that, in the Mekong Delta, so much sand has been mined that the riverbanks are now collapsing. About half a million people will have to be relocated away from this mighty river.

In India, sand in rivers acted as part of the local groundwater system. But when the sand was taken away, the overall water table dropped – not just in the river, but for kilometres around. As a result, many rivers no longer even run with water in the dry season. Which means, of course, that the farmers can't grow crops anymore. And along several rivers, major bridges are now in danger of collapse, because their foundations are no longer secure, due to the sand being physically removed from around them.

In Indonesia, over two dozen small islands have disappeared entirely, having been mined to below sea level. International boundaries have even had to be redrawn.

In the oceans, dredging is as brutal as a sledgehammer – it sucks up everything, whether it's animal, vegetable or mineral. Dredging marine sand is catastrophic for local sea life. It simply sucks the living creatures into the pumps, and then they ain't alive no more.

And in addition, the massive plumes of sediment and increased turbidity that spread out for kilometres drastically interfere with marine life. It's like a choking sandstorm, but underwater. Think of

TIME IS RUNNING OUT FOR SAND

Effects of ocean dredging

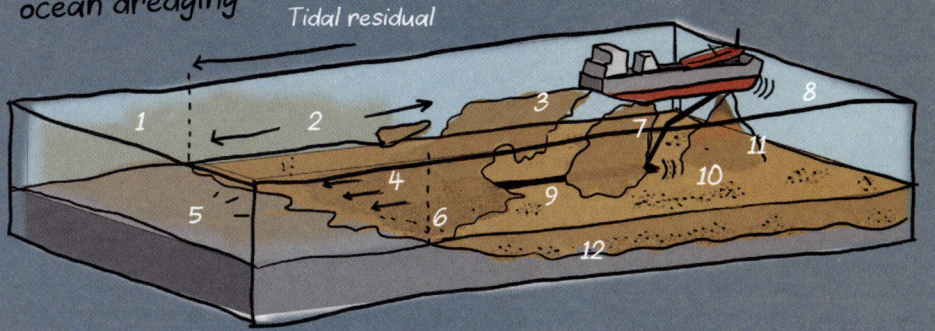

Tidal residual

1. Increased turbidity 2. Far field changes in tides and currents 3. 'Passive' sediment plume 4. Plume dispersal 5. Seabed sediment veneers 6. Deposition from sediment plumes 7. 'Active' overflow plume 8. Ship/machinery noise 9. Seabed removal: depth change 10. Draghead noise 11. 'Active' screening plume 12. Base of deposit

Water column

Resource – sand and gravel

Bedrock

how bad air pollution can be for the health of humans – and then multiply that by many thousands.

The effects of runaway sand mining include loss of land and biodiversity. But there are also major changes in water flow and water supply in the landscape – such as coastal erosion and pollution of rivers.

We've known since the mid-1970s that Climate Change is causing an increase in extreme weather events. In some cases, the mining of sand aggravates these events, because we have lost the natural protection sand provides against floods, droughts and storm surges.

WE CAN FIX IT

We can do so much better.

We can use alternatives to sand, such as incinerator ash, crushed rock, industrial slag and waste, and, in some circumstances, recycled plastic. We can reuse sand-based materials (such as clean demolition waste and concrete) instead of dumping them into landfill. We can use clean rubble as a base aggregate for the foundations of roads and buildings.

We can think outside the box, finding new methods for building structures, and shifting to more efficient materials. But at the moment, lack of awareness of the global shortage of sand is the major problem. The issue is not even on the radar of most political leaders around the world.

We need to curb our appetite for this over-mined grain before the sands of time run out.

Sand, concrete & greenhouse gases

Cement is the 'glue' that holds various ingredients (sand, gravel, etc) together to make concrete. In terms of global warming, the cement industry is one of the larger producers of carbon dioxide, accounting for about 5% of global emissions. When you make one tonne of cement, you generate 0.9 tonnes of carbon dioxide.

Today, most cement is made by heating calcium carbonate to make quicklime (CaO) – an important part of the 'glue':

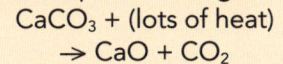

$$CaCO_3 + \text{(lots of heat)}$$
$$\rightarrow CaO + CO_2$$

A bit more than half of the emissions come from the actual chemical process of making quicklime, which releases CO_2 into the atmosphere. The rest comes from burning carbon (coal, etc) to generate the necessary high temperatures to make this chemical process happen.

However, it is entirely possible to make concrete without burning any carbon at all.

With modern chemistry, you can totally ditch quicklime. One very promising alternative is stuff called 'geopolymers'. Switching from quicklime to geopolymers would remove about half the greenhouse-gas emissions (the CO_2 on the right side of the equation). You will still need heat for making the ingredients, but that can come from renewable sources of energy.

The technology is already here. Let's use it.

VEGAN DIET

Way back in 1924, a 14-year-old kid called Donald Watson saw a pig being slaughtered. The pig was terrified and screaming. Donald was equally upset. Even though Donald was a farm kid, he was so moved by the slaughter of the pig that he stopped eating meat. He later stopped eating dairy as well.

A few decades later, in 1944, the very same Donald invented the word 'vegan' – by joining together the first and last syllables of the word '*vege*tarian'.

VEGAN AND HEALTHY?

People sometimes wonder if you can be truly healthy on a diet that excludes both meat and dairy. The answer is definitely a strong *yes*. But you need to think about it! You have to understand your food much more deeply than somebody who eats the standard meat and three veg (one of each colour – white, green and yellow/orange).

There are many really good reasons for changing to a plant-based diet. Some include concerns about animal suffering and cruelty, or your own health, or the environment – and even fatuous reasons, like imitating your favourite celebrity.

Ethical (or moral) vegans are against using animals for any reason; *environmental* vegans are especially concerned that our current industrial-grade farming practices are both unsustainable in the long term, and damaging to the environment in the short term. From a health point of view, plant-based diets have been linked to lower risks of obesity and many chronic diseases, such as type 2 diabetes, heart disease, inflammation and cancer. And evidence links colorectal cancer with eating lots of red and processed meats. But the health benefits of going vegan don't come without some thought.

Clare Collins, Professor in Nutrition and Dietetics at the University of Newcastle, says there are four essential nutrients that you must be especially conscious of if you choose to go vegan. If you stop eating meat or dairy products, you'll usually struggle to get a decent supply of these micronutrients: vitamin B_{12}, calcium, iodine and iron.

Australia & veganism

Between mid-2015 and mid-2016, Australians did more Google searches for the word 'vegan' than any other nationality. Australia is the third-fastest growing vegan market in the world, following China and the United Arab Emirates.

Do herbivores eat meat?

If a herbivorous animal was hungry, could it survive by eating meat?

Sometimes.

The traditional belief is that carnivores eat meat only (teeth for ripping, short gut) and herbivores eat plants only (teeth for grinding, long gut, sometimes with specialised fermentation chambers). In between them are a few oddball species like us humans, who eat plants and meat (teeth that can rip or grind, medium length gut).

However, many herbivorous animals have been seen eating the placenta after giving birth – and placenta is made from meat.

Deer (classic herbivores, with fermentation chambers in the gut) have been seen eating bones, dead fish that have drifted to shore (at the rate of eight per minute), the guts of other (dead) deer, and entire birds. Other supposed herbivores, such as camels, giraffes, pigs, cows and sheep, have been seen eating other animals, or animal parts left over from a carnivore kill. Pandas (which normally eat only a specific species of bamboo) have been seen eating peacocks.

In *Seven Pillars of Wisdom*, T.E. Lawrence writes about racing camels being fed boiled meat to improve their performance.

Maybe the herbivores are deficient in a specific nutrient? Or maybe they are clever enough to not pass up a free meal?

VITAMIN B$_{12}$

Let's start with vitamin B$_{12}$ (also called 'cobalamin'). It's essential for making DNA, fatty acids, red blood cells, the myelin sheaths around your nerves, and some neurotransmitters in the brain.

A deficiency of B$_{12}$ can cause a fast heart rate or palpitations, bleeding gums, bowel or bladder changes, a sore tongue, impotence, tiredness, weakness, infertility, light-headedness, and a low tolerance for exercise. Certainly, if you're feeling tired and weak, you won't feel like pumping iron. A severe deficiency can cause reduced heart function and serious neurological symptoms (memory problems, psychosis, etc). Sometimes, the neural damage is irreversible. Children can suffer poor development and growth, and difficulties in movement.

In the US and UK, B$_{12}$ deficiency is found in about 6% of those under 60, but 20% of those over 60. It can reach 40% in Latin America, and 80% in parts of Asia and Africa.

There are three main pathways to B$_{12}$ deficiency. First is decreased intake – such as in vegans, vegetarians and the malnourished.

Second is decreased absorption of B$_{12}$ from the gut, even though enough is present in the diet. This lack of absorption can be caused by pernicious anaemia, some gut conditions (surgical removal of the stomach, chronic pancreatic inflammation, intestinal parasites, etc), nitrous oxide consumption, and some regular medications.

Third is an increased need for B$_{12}$, such as in AIDS, and problems with the life span of red blood cells.

On average, if you are in good health, you carry about 3–5 years' supply of B$_{12}$ in your body (about 2–5 mg). So it can take a while for a B$_{12}$ deficiency to show itself. About half of your B$_{12}$ is stored in your liver.

Vitamin B$_{12}$ is not directly made by any animal, plant or fungus. However, it is found in animal-based foods, such as meat, milk and other dairy products (i.e. from ruminants such as cows and sheep).

So how does B$_{12}$ get into the flesh and milk of cows and sheep? Once again, bacteria are our friends. Certain bacteria colonised the long and complex gut of these ruminants, and these bacteria made the B$_{12}$, which then spread throughout the entire body of the animal.

Vegans can get a little vitamin B$_{12}$ in some algae and plants that have been either directly exposed to bacteria, or indirectly exposed via soil or insects. It's also found in some mushrooms or fermented soybeans – but only traces.

So vegans really need to consume foods with added vitamin B$_{12}$, like fortified non-dairy milks, or take B$_{12}$ supplements.

CALCIUM

Calcium is essential for good bone health – as well as for proper functioning of the heart, muscles and nerves. If calcium levels are too low, you can get osteoporosis. Your bones become less flexible and more brittle, making them more likely to break.

Calcium intake is deeply connected to the concept of peak bone mass. It turns out that you reach the maximum amount or mass of bone in your whole life in your early 20s if you are female, or your late 20s if you are male. So if you don't keep your calcium intake high before this, you will reach a lower amount or mass of bone. And you can never go higher than this level. All you can do is maintain your mass of bone at this level, or let it drop. So getting enough calcium before you reach your early/late 20s is essential.

Calcium is abundant in milk and milk-based foods. Vegans can get calcium from tofu, some non-dairy milks with added calcium, as well as nuts, legumes, seeds and some breakfast cereals.

Surprisingly, both vegans and vegetarians usually need to eat more calcium than meat eaters. That's because some plant foods contain chemicals that unfortunately reduce the absorption of calcium into your body – and vegetarians and vegans usually eat more of these plants. These calcium-blocking chemicals include oxalic acid (found in spinach, rhubarb and beans) and phytic acid (found in soy, seeds, grains, nuts and some raw beans).

IODINE

Surprisingly, vegans can also be deficient in iodine.

Iodine is essential for making thyroid hormones, and for the developing central nervous system. The symptoms of iodine deficiency can include feeling cold, weight gain, tiredness and weakness, constipation and poor mental function. Women with low iodine levels are more at risk of miscarriage and stillbirth, and the baby can suffer mental retardation (because iodine is essential for full brain development).

Vegans don't eat the usual sources of iodine – seafood, dairy products and eggs. However, if the local soil is rich in iodine (e.g. by being close to the ocean), the plants grown in it will often carry enough iodine. (And yes, inland soils can sometimes be deficient in iodine.) Vegans can also eat seaweed, and foods that have added iodine such as salt, some breads and some non-dairy milks.

Bacteria and B_{12}

To make B_{12}, bacteria need the trace elements of copper and cobalt. The soil in which the grass grows that the ruminants graze on has to carry these metals, as they are essential for the synthesis of the molecules of B_{12}. The volcanic soils (and volcanic soils are usually very rich in all kinds of nutrients) on the North Island Volcanic Plateau in New Zealand were unfortunately low in cobalt. In the early 20th century, cattle grazing on these volcanic soils suffered from 'bush sickness'. In 1934, scientists worked out that the volcanic soil had virtually no cobalt, leading to the cattle suffering a B_{12} deficiency.

In a similar case, in the 1930s, sheep raised on the coastal sand dunes of South Australia suffered from 'coast disease'. Again, this was due to a B_{12} deficiency, caused by a lack of both copper and cobalt in the soils. This was fixed by feeding the sheep 'cobalt bullets' – a dense mix of clay and cobalt oxide.

So why would vegans be prone to iodine deficiency? Well, swallowing iodine is only half the battle. Just like with calcium, some foods can reduce your absorption of iodine. If you love your brassicas – foods like cabbage, broccoli, and brussels sprouts – and sweet potato and corn/maize, you're also getting a dose of chemicals in these vegetables that can block absorption of iodine. The end result is interference with the production of the thyroid hormones.

IRON

Finally, we come to iron. Most people know that getting enough iron can be problematic on a vegetarian or vegan diet. Iron is essential to make the haemoglobin in red blood cells (which carry oxygen around your body), and myoglobin (which can store oxygen in your muscle cells, and which is related to the red colour of some meats). So a deficiency can leave you feeling tired, and even having a weakened immune system.

It is easy to get enough iron if you eat wholegrain cereals, meat, chicken and fish. In Australia and New Zealand, most people get their iron from these sources.

There *is* iron in some plants – such as lentils, beans, tofu, cashew nuts, pumpkin seeds, kale and more. Unfortunately, it's non-haem iron, which your body can't absorb as well as it absorbs the haem iron from meat.

You can boost your absorption of iron by eating vegetables and fruit that are rich in vitamin C at the same time as eating the plants containing iron. Fruits rich in vitamin C include mango, citrus fruits, kiwi fruit, pineapple, and so on. Vegetables high in vitamin C include capsicum (probably the King of vit C), broccoli and other brassica vegetables, tomatoes, potatoes, and so on. The vitamin C converts the non-haem iron into a form that is better absorbed. (The chemistry is that the vitamin C reduces the iron from the ferric (2+) form to the ferrous (3+) form.)

And if you're vegan, don't have a cuppa to finish off your meal!

Black tea contains chemicals that can reduce your absorption of iron! There are also other foods that can reduce the absorption of iron (both haem and non-haem), such as soy products, wine, coffee and more. If you are at risk of low iron levels, see a professional dietitian.

DEEPER UNDERSTANDING

Vegans need to understand food much more deeply than meat-eaters – and that is good.

However, if you've been a vegan for a long time, there are more nutrients to keep your eye on. You also need to watch your levels of vitamin D, omega-3 fats and protein.

Finally, vegans must take an extra level of care with their diet plans if they are pregnant or breastfeeding, or bringing up their children as vegans. You have to juggle a bunch of potentially conflicting needs – getting enough kilojoules, not dropping your intake of the right kinds of fats, and also keeping up an adequate intake of necessary nutrients. There are macronutrients (fats, proteins, carbohydrates) and micronutrients (everything else, including vitamins, trace elements, etc). Again, it's extremely worthwhile getting the advice of a professional dietitian.

With sufficient background knowledge, it's totally correct to say that you can be vegan, be healthy, give birth to healthy children and raise them to adulthood as vegan. You just have to choose the right food options.

When it comes to healthy eating for vegans, it ain't the meat *or* the motion – it's the micronutrients!

Yes, it takes a bit of effort to get all your nutrients from a vegan diet. But it's not as though eating meat and animal products is a guarantee of healthy eating!

FISHER SPACE PEN – OUT OF THIS WORLD!

When it comes to the Fisher Space Pen, we're talking about something close to my heart. But for the non-obsessed, knowledge of the Fisher Space Pen boils down to three main scenarios.

First, you've never even heard of a Fisher Space Pen. That's zero knowledge.

Second, you came across it in that famous 1998 episode of *Seinfeld*, 'The Pen': 'This is an astronaut pen. It writes upside down. They use this in space.' If you haven't come across *Seinfeld* yet, you will one day.

Third, somebody told you (or you read on Facebook) that NASA spent billions of dollars to develop a special ballpoint pen that would work in space, but the canny Russians simply used a pencil.

Well, as I'm sure you suspected, the third scenario is pretty loose with the truth. Actually, it's ludicrously incorrect. (In reality, instead of having spent billions of dollars, NASA actually spent a measly $2.39 per pen – yup, less money than what you'd spend on a single cup of coffee.) But in my experience, this myth is the most common memory that people have about the Fisher Space Pen.

SPACE, THE HOSTILE FRONTIER

Space travel is dangerous. A spacecraft is a closed environment – like a submarine running underwater, but with extra hazards. In 1967, the *Apollo 1* command module caught on fire, killing all three astronauts, due to a fatal combination of combustible materials, a spark and a high-oxygen environment.

In space, the lack of effective gravity means that everything keeps on doing what it was doing when you let go of it. Solid stuff doesn't automatically drop to the floor. A warm gas (like the air you breathe out) doesn't quickly spread out in all directions.

When an astronaut sleeps, they need a little fan blowing onto their face. Otherwise, the carbon dioxide they breathe out would just hang around their face. When carbon dioxide reaches 5% levels in the air you breathe, it starts becoming toxic.

The ink from a ballpoint pen could also cause a risk. It might have nasty toxic fumes, which would just hang around in a spacecraft and be potentially dangerous to astronauts.

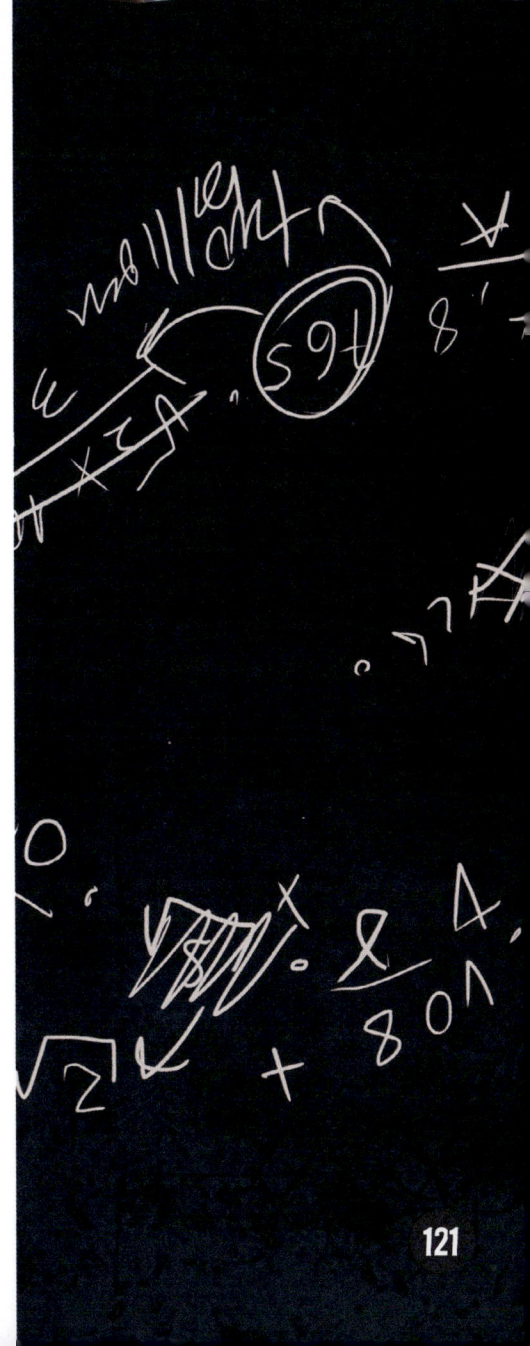

Fisher Space Pen & medicine

When I started as a very junior medical doctor, part of my job was to fill out lots and lots of paper documents for the senior doctors – usually for some kind of investigation request. This was mostly for blood tests, but sometimes included X-rays and ultrasounds.

The ward rounds were very busy, and very speedy. So I would have to fill out the forms on the nearest wall, above my head. Thanks to my engineering background, I knew of the Fisher Space Pen and its ability to write uphill, so I had one of my own. Bingo, problem solved!

Using a pencil in the absence of gravity raises different problems. If the tip of the graphite point broke off, it would just float in the air. So would the wood shavings or graphite dust left over from sharpening a pencil. This floating debris could land in your lungs or your eyes.

But this debris can have worse consequences. If it conducts electricity (and graphite does) it might disrupt the operation of electronic components and electric switches. Graphite is also flammable.

The Russians had problems with pencils. The cosmonaut Anatoly Solovyev said, 'Pencil lead breaks ... and is not good in space capsule: very dangerous to have metal lead particles in zero gravity.'

ASTRONAUTS NEED TO WRITE

While we have launched lots of spacecraft since the 1960s, even today spacecraft are not mass-produced items. Each one is hand-built, and is an experimental vehicle. If it carries people, the stakes are much higher. For each stage of a spacecraft's flight, astronauts keep checklists and other permanent, detailed documentation to help avoid mistakes, and for future reference.

So astronauts in a spacecraft have to be able to write down what happens – even today. Nowadays, the 'writing' is mostly electronic, but this was not the case in the past.

In 1965, NASA spent $4,382.50 on 34 mechanical pencils. That works out to be $128.89 per pencil. There was a public outcry at the perceived waste of funds. While each mechanical pencil had cost just $1.75 to buy in the shop, NASA spent the rest of the money on the necessary modifications for using it in space.

Back then, space travel was so risky that the astronauts wore a full space suit, including gloves, for the entire duration of each mission. Writing with your bare hand is very different from writing while wearing a very stiff glove. NASA had to individually modify each mechanical pencil with lightweight, high-strength materials, so that it could be used while the astronaut was wearing bulky spacesuit gloves. This was expensive. Anybody who has tried to build anything will understand how much time it can take to make even a seemingly simple contraption.

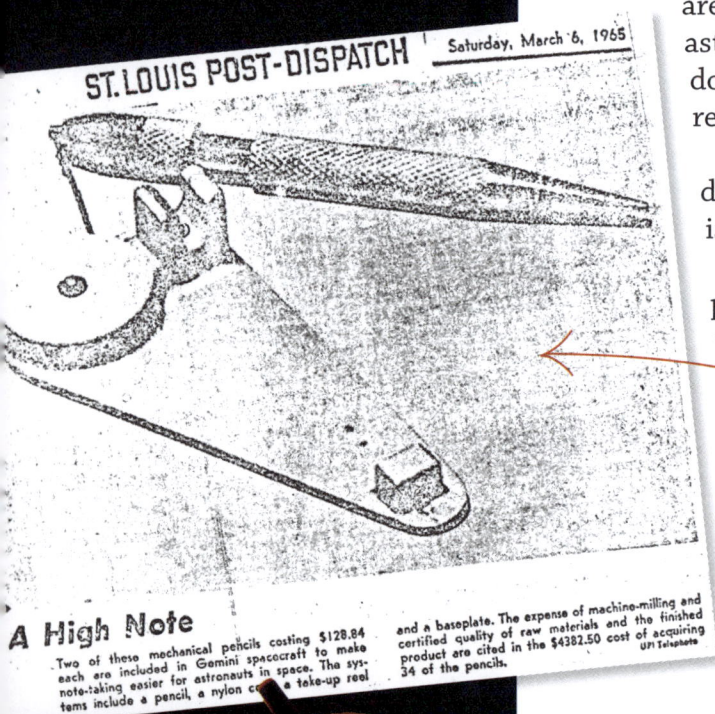

ST. LOUIS POST-DISPATCH   Saturday, March 6, 1965

A High Note

Two of these mechanical pencils costing $128.84 each are included in Gemini spacecraft to make note-taking easier for astronauts in space. The systems include a pencil, a nylon cord, a take-up reel and a baseplate. The expense of machine-milling and certified quality of raw materials and the finished product are cited in the $4382.50 cost of acquiring 34 of the pencils.

ENTER THE FISHER SPACE PEN

By a coincidence, in that same year of 1965, Paul C. Fisher from the Fisher Pen Company in California independently developed and then patented the AG-7 pen. (AG stood for 'anti-gravity'.)

The AG-7 was made from solid brass, which was plated with a hard, shiny chrome coating. You pressed down on the top of it to extend the replaceable ink cartridge (or refill), and then pressed a button on the side to retract the cartridge. At the very tip of the cartridge was the ball that carried the ink onto the paper (see Fig. 5 in the US Patent documentation). It was made of very high-quality tungsten carbide, and it sat inside a high-precision socket of stainless steel.

But the inside of the ink cartridge was where the magical science happened.

Even today, all ballpoint pens (except for the Fisher Space Pen) rely on gravity to make the ink come out. But in the Fisher Space Pen, even the 'ink' was not a regular ink. The ink was (and here comes a special word) 'thixotropic'. Thixotropic means being able to change from a semi-solid gel to a liquid, by adding pressure.

Tomato sauce and ketchup have this strange property of being 'thixotropic'. The tomato sauce sits inside the bottle as a gel, not a liquid. But if you 'attack' the tomato sauce by shaking the bottle, the tomato sauce changes temporarily from a gel into a slow-moving liquid.

If you want to get technical, thixotropy is the property called 'time-dependent shear-thinning'. But in plain English, when you disturb a thixotropic gel, it temporarily becomes thinner and more liquid. So when you shake the tomato sauce bottle, for a little while the tomato sauce becomes thin enough to drip onto your food.

So, the ink inside the Fisher Space Pen cartridge is thixotropic – it's a thick gel. When you start writing, the tungsten carbide ball rolls across the paper. At the same time, it rubs up against the gelatinous ink, and shears the chemical polymer bonds in the gel. The gel turns into a liquid, coats the rolling ball and is promptly deposited onto the paper.

This works *only* if the gel is in intimate contact with the tungsten carbide ball. Unfortunately, gravity is too weak to keep the gel hard up against the surface of the ball.

For intimate contact, you need to push that special ink/gel with a bit of pressure. Nitrogen gas at a pressure of over two atmospheres (fairly close to the pressure in your car tyres) is perfect for the job –

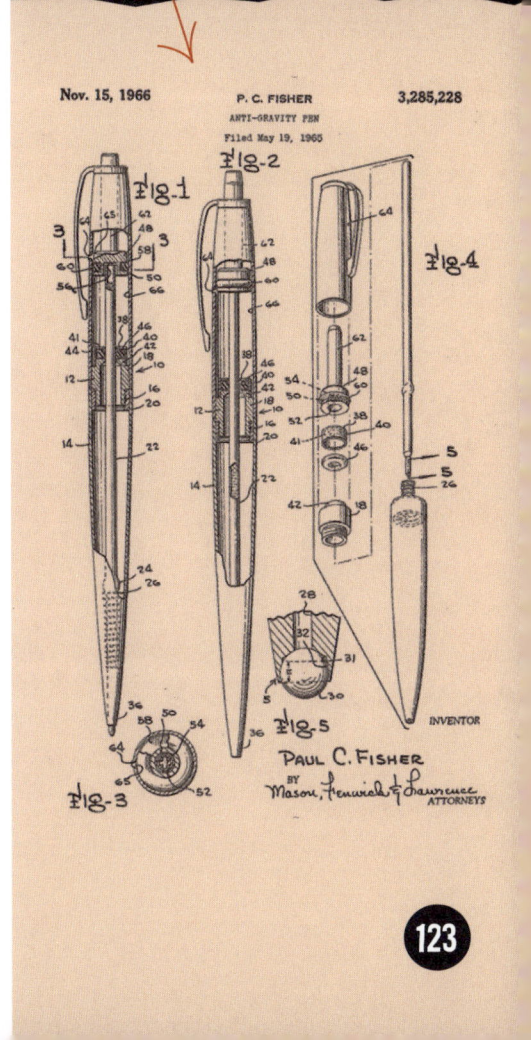

SHAKE WELL!

Writing

Nov. 15, 1966   P. C. FISHER   3,285,228

ANTI-GRAVITY PEN

Filed May 19, 1965

Fig. 1   Fig. 2   Fig. 4

Fig. 3   Fig. 5

INVENTOR

PAUL C. FISHER

BY Mason, Fenwick & Lawrence
ATTORNEYS

and because nitrogen is an inert gas, it does NOT chemically react with the ink. That's why, immediately above the thixotropic ink, the pen has a reservoir of pressurised nitrogen gas, which is separated from the ink by a sliding float. This pressurised and highly engineered pen would write in a weightless environment, underneath water and other liquids, and in temperatures ranging from -35°C up to +120°C.

There was only one small problem – the ink would turn green at high temperatures. But at least you could still write and your notes could still be read …

Inventing a colourful Fisher Space Pen

I love Fisher Space Pens, and I also love colour. The problem is that the Space Pen came only with a chrome finish.

So I bought a whole bunch of non-Fisher brightly coloured pens, removed their standard internal cartridges and inserted super-duper Fisher Space Pen cartridges (which I bought separately). They worked perfectly and looked great.

At that time, we had little kids. Somehow, my beautifully coloured pens with their magnificent Space Pen cartridges began to vanish, until there were none left.

A few years later, I was doing some digging in our garden – there were the pens! And after a bit of cleaning, the cartridges worked again. I was so happy.

The lesson from this?

Little kids have an amazing ability to make anything vanish.

SO HOW MUCH DID THE SPACE PEN REALLY COST?

The total cost of developing this technology was claimed to be around US$1 million. It was *not* paid for by NASA. The cost was carried entirely by the private Fisher Pen Company (so that's completely different from saying NASA spent billions of taxpayers' dollars).

Paul Fisher tried desperately to get NASA to officially endorse his space pen, which would have been great for marketing purposes. But NASA, as a government agency, wouldn't promote any specific commercial product. NASA could, however, recognise a good piece of kit (a pen that was both safe to use and fully functional in space) and so they bought it.

By 1967, NASA had ordered 400 Fisher Space Pens for their Apollo program. The Russians followed suit, ordering 100 pens and 1,000 ink cartridges to use on their Soyuz space missions. In other words, as soon as the Russians could stop using pencils, they did. And because both NASA and the Soviet space agency were buying in bulk, they apparently each got the same 40% discount. Instead of paying $3.98 for each pen, they paid $2.39.

So the tale of the billion-dollar NASA pen vs the Russian pencil has an element of truth – but not enough to get in the way of a good story.

And the AG-7 Space Pen? It's still delivering a payload of income for Fisher today – without that 40% discount.

Fisher Space Pen & water

Yup, I still have my favourite pens with Fisher Space Pen cartridges. But I have since found some magic paper that is waterproof. Of course, I tested it (you always have to do the experiment). I found that I could write with a Fisher Space Pen on the waterproof paper, while water is pouring across it, while holding it above my head.

One day (and I hope that day comes soon), I will be showering and have a brilliant thought. All I will have to do is reach out of the shower, grab my Fisher Space Pen and waterproof paper, make a note and I'll be on my way to a Nobel Prize – or, at least, an Ig Nobel Prize.

FISH
EXERCISING

I used to read a book to my kids when they were little – *Enzo the Wonderfish*. It was a beautifully illustrated and funny book about a kid who desperately wanted a pet, but her crummy parents would only get her a fish. So when she wanted to exercise her fish, she had to carry it inside the fishbowl all around the park. She was always worried about the water being too rough and making her fish seasick.

It turns out that exercising fish is a thing for some scientists.

Hard to believe, but tell me more.

In fact, these scientists exercise their fish with a coffee plunger (and a magnet)!

Seriously, how do you exercise a fish with a coffee plunger and – I'll bite – why?

WHY EXERCISE FISH?

Why on earth would anybody in their right mind exercise a fish?

The answer? To learn more about humans.

In humans, fitness, obesity and metabolism are all really complicated and interlinked – which is why we are still trying to work it all out.

As the 'obvious' next step in trying to understand humans, scientists are increasingly using zebrafish as a 'vertebrate model organism'. It turns out that we humans have more genes in common with the zebrafish than we do with the 'classic' research organisms – the fruit fly, and the nematode *C. elegans*. In fact, humans and zebrafish have the same genes involved in some cancers and developmental disorders. Scientists doing this work on human fitness come from fields as varied as biomedicine, developmental biology, evolutionary biology and behavioural science, covering conditions as varied as leukaemia, diabetes, retinal damage and skin cancer.

So zebrafish are appearing not just in a single little sub-sub-subgroup of scientific/medical knowledge – they're spreading across the board. (Yes, folks, take up science as a career – you can never predict what weird fields of research you might end up in!)

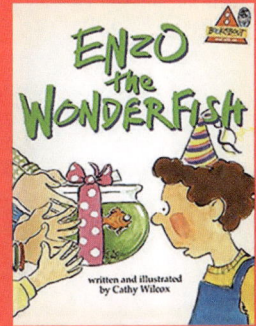

Part of the deal is that if scientists want to exercise fish, they first have to know how fit they are to start with. You do these same sort of baseline fitness tests with different people. After all, if you wanted to compare different people at running, you would want to know if you were putting a couch potato up against a marathon runner. In fish science talk, this is technically 'assaying swimming performance'.

One way to exercise zebrafish is to manually chase them, until they're exhausted. But you can do this only for a few fish at the same time, and while you're taking your fish for their equivalent of a walkie, you can't do anything else (like make morning coffee).

There are other ways to exercise fish. One is the traditional Brett-type swimming chamber, in which individual fish swim against water currents. But these devices are expensive, can exercise only a few fish each day and take up lots of room.

DIFFERENT KIND OF COFFEE 'RUN'

So our multinational team of scientists (Australian, Japanese, Canadian and Polish), led by Professor Shinichi Nakagawa from the University of New South Wales and the Garvan Institute of Medical Research, wanted a way to cheaply and consistently exercise a bunch of fish. They came up with the brilliant idea of using a French coffee press (or coffee plunger to Australians) with a magnetic stirrer, to create a circular flow of water. This is how the scientists exercised a lot of fish, quickly, consistently and cheaply.

They started with a coffee plunger (okay, if you must know, it was the IKEA Upphetta coffee/tea maker, 400-ml volume, 8-cm diameter), and gently placed a 'rotating magnetic stir bar' on the bottom. They controlled the magnet by placing the coffee plunger on top of a Single-Plate Labtek Magnetic Stirrer. They filled the glass coffee pot with 300 ml of water, inserted the plunger, and added the zebrafish *above* the plunger (so that they couldn't run into the spinning magnet). Then, they turned on the stirrer and gently increased the rotating speed. Once they saw that the single-plate device did what they wanted, and didn't hurt their zebrafish, they upgraded to the Labtek 10 Place Stirrer.

Voilà! Cheap as chips, consistent, and ten fish exercised at the same time.

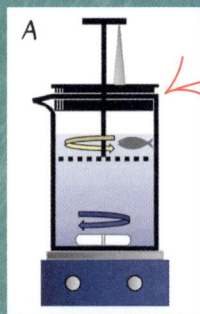

IT WORKS, BUT ...

There had been previous attempts to train and test zebrafish against a spinning water flow. But these had created a large central water vortex, which moved randomly. So the results were very inconsistent.

Being good scientists, the team tested their system. They certainly didn't want to harm their precious zebrafish, and they also wanted clear results. They tested for consistency, whether the zebrafish could swim faster after training, weight changes, and so on.

First, they were getting consistent results (hoorah!).

Second, the maximum swimming speed significantly increased in the exercise-trained group, but was unchanged in the control group. This makes sense, and is exactly what we see in humans.

Third, there was a weak link between the gender and the length of the fish, and its maximum speed.

And finally, the exercise-trained fish did not lose weight. But as you might expect, while this coffee-plunger training was a great leap forward, there were some 'limitations' to this study.

First, the water spins more slowly at the centre of the coffee plunger than out near the rim. This could mean that heavier fish could lose control of their movements at lower speeds than if they were tested in a linear flow, such as in a Brett-type swimming chamber. After all, riding a bicycle in small circles is more difficult at low speeds, where there is little momentum.

Second, the fish might get exhausted sooner in a spinning water flow than in a linear-flow swimming chamber. With a linear flow, they use all their muscles to swim. But if a fish swims in a circle, it might be using the muscles on one side more than on the other. So theoretically, to quote the research paper, 'imbalanced use of one side of the fish musculature due to circular swimming may result in premature, rather than complete, exhaustion'. One solution could be to have them swim in both clockwise and anticlockwise directions. This might work if your laboratory magnetic stirrer would rotate both ways, but not all of them do this.

And third, swimming in a circle is rather artificial. After all, fish in the real world don't usually spend a lot of time swimming in an 8-cm circle. And not many human exercises are carried out in a very small diameter circle (apart from the Olympic sports of hammer throw, discus and shot put). But you have to admit it was very creative to make a gym for zebrafish out of a coffee plunger.

Maybe aquarobics instructors need a bigger version of the coffee plunger to get maximum effort out of their pool pals?

Sharks & the Brett-type swimming chamber

Think about sharks. They always seem to be swimming. But do they have to?

Well, some do, and some don't.

The ones that have to swim are called 'obligate ram ventilators'. They force water (containing essential oxygen) to move over their gills by holding their mouth open and swimming. If they want to 'breathe', they have to swim. These include great white, mako and whale sharks.

The sharks that don't have to swim to breathe can get enough water to move over their gills by 'buccal pumping'. They close their mouth, squeeze their cheeks (or 'bucca' in Latin), and the water flows over their gills. These include the nurse and bullhead sharks.

One way to study this is with a 'closed swim tunnel'. It's basically just a transparent pipe big enough for your fish or shark, with water flowing through it. Think of it as a treadmill, but for fish. You can control the water – its speed, temperature, cleanliness, oxygen level, etc. You can also take lots of measurements of the fish – its speed, muscle temperature, heart rate, tail flapping, etc.

J.R. Brett did a lot of the early work in this area in the 1960s. Hence the name Brett-type swimming chamber.

SURFING SPACE-TIME – EINSTEIN'S RIGHT AGAIN!

For my money, 'Bizarre cosmic dance offers fresh test for General Relativity' is a great headline. 'General Relativity' is physics talk for 'Gravity', so testing our major theory of gravity with a bizarre cosmic dance floats my boat!

Einstein and General Relativity are inseparable. I love how General Relativity keeps on passing every test we find for it. But I confess there's a tiny part of me that wants Einstein to be proven wrong! (Trust me, it's not Tall Poppy Syndrome, it's for a good cause.)

But our bizarre cosmic dancers need Einstein's version of Space-Time to surf on. So we won't be dumping Einstein just yet …

DIMENSIONS OF SPACE-TIME

Most of us have a rough idea of 'dimensions' – and I don't mean dimensions as in 'size'. I mean dimensions as in 'a painting is two-dimensional, while reality is three-dimensional'.

The three dimensions of reality are left-right, backwards-forwards and up-down (the mathematicians call them x, y and z). These dimensions define a specific location, e.g. 'Let's meet on the corner of Einstein and Beethoven streets, on the first floor of the Tennis Building.' But then you need another dimension, time, to specify *when* this event will happen – obviously 10.30 am for a nice cuppa. Put these four dimensions together and, bingo – Space-Time.

Now for a few comments …

First, Einstein started us down this pathway in 1905. But it was the mathematician Hermann Minkowski, in 1908, who showed how to write Einstein's equations in a unified way, with space and time on an equal footing. So now we know that these four dimensions are all 'married' to each other. Time is not a completely unlinked, separate and independent dimension.

Second, for three of the Four Forces in our Universe (Electromagnetic Force, Strong Nuclear Force and Weak Nuclear Force), events happen on the background of Space-Time. But the fourth Force (Gravitational) is a 'bending' or distortion of Space-Time.

The physicist John Wheeler said, 'Mass tells Space-Time how to curve, and then curved Space-Time tells mass where to go.' So Gravity is just 'curved Space-Time'.

Let me explain with the example of a flat trampoline sheet. We'll pretend that it is flat Space-Time – like you get in empty space, far away from any source of gravitation.

Put a golf ball on the trampoline, and flick it with your finger. This is equivalent to a comet passing through our flat, undistorted Space-Time. The golf ball/comet travels in a straight line.

Size of stars

The size of a star depends on the balance of two forces, which vary during the evolution and life of that star. One force is gravity, always sucking towards the centre. That depends on its mass, which can change during the star's evolution. The other is the awesome force of its nuclear burning pushing outwards, which also changes. Our Sun is about 1.4 million km across.

In some cases, when the star is in its final stages of burning hydrogen atoms in its core via nuclear fusion, it can expand mightily. The classic example is a red giant. These start off as stars with a mass in the range of 0.3 to 8 times the mass of our Sun, and in their later stages of evolution can expand to be as much as 200 times larger.

But when the nuclear fuel runs low, gravity wins and the star shrinks. At this stage, it might shrink down to a white dwarf star (about 13,000 km across), or a neutron star (20 km across), or a black hole (no size at all).

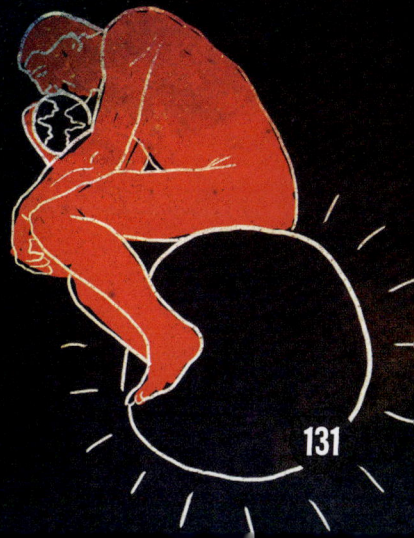

Rotating more quickly?

Think of the classic ice-skater pose, rotating while standing on just one ice skate – with their arms stretched outwards. They then bring their arms close to their body – and they speed up. (It's related to Conservation/ Transfer of Angular Momentum – check it out on Wikipedia.)

The same happens with white dwarfs. In about 5 billion years, our Sun will first expand and throw off some mass, but then shrink down to a white dwarf some 100 times smaller, but still keep 80% of its initial mass, making it spin a lot faster. The fastest rotation time we have found for a white dwarf that is all alone is about 317 seconds (5 minutes and 17 seconds).

But if the white dwarf star is really close to another star, it will gravitationally 'suck' mass from this other star onto itself. This accumulation of mass, close in towards the rotational axis of the white dwarf, will speed it up (just like the ice skater pulling their arms in close to their body and speeding up).

Now put a heavy bowling ball in the middle of the trampoline – let's pretend this is a heavy star. It distorts the previously flat trampoline/Space-Time, and then sits in the little curved 'well' that it created.

Again, flick the golf ball/comet roughly in the direction of the bowling ball/star. This time, the golf ball/comet travels in a curved path. Depending on its speed and direction, it will curve gently, or curve a lot, or even end up dropping into the well and ramming the bowling ball/star.

So mass (bowling ball/star) curves Space-Time (the trampoline sheet), and curved Space-Time (the well in the trampoline sheet) tells the mass (golf ball/comet) what path to take.

In our Solar System, the well in Space-Time put there by the Sun sits in the middle of the Solar System. But the well does *rotate*, because the Sun rotates (about once a month). And planets that orbit around the Sun carry their much smaller wells in Space-Time with them.

BIZARRE COSMIC DANCE

Yes indeed, the cosmic dance in this story is truly bizarre.

It's happening about 15,000 light years away from us (though that's still well inside our Milky Way galaxy). Two really weird stars, which originally started as regular stars, are orbiting around each other crazily fast. Their official, and boring, name is PSR J1141-6545, and they're called a 'binary radio pulsar'.

Space is big, and the distances are huge, so it can be hard to get a grip on the scale of things. It helps if we think about these objects in terms of the Earth and the Sun.

So, how close are these two dance partners to each other? *Very* close – only about the diameter of our Sun, or about 1.4 million kilometres apart.

And their speed? Well, the Earth orbits the Sun in one year, but these objects are orbiting each other in five hours! So while the Earth tears around the Sun at a cracking 30 km per second, these stars are moving almost ten times that fast – at about 280 kilometres per second!

Now, both of these weird objects used to be a regular star similar to our Sun, but more massive. And after quite different expansion phases, they each ran low on fuel and were each finally crushed by their own gravity down to a fraction of their former size.

One of them ended up as a white dwarf. Unusually, it was the first to evolve from a regular star. (Normally, when there's a pair of stars orbiting their common centre of gravity, if one of them evolves into a white dwarf, this event happens after the evolution of the other star.) It went through its life cycle, and then shrank down to about 100 times smaller than our Sun – roughly the size of the Earth, or about 13,000 kilometres across. But while the Earth rotates once every day, this white dwarf was rotating once every 200 seconds! You can think of a white dwarf as the burnt-out core of a star, or a stellar remnant.

The reason it could spin so fast was that, in its younger days, this white dwarf had had a regular star as its companion. In the beginning, when they were both regular stars, they had a nice balanced relationship. But things changed when it became a white dwarf – because white dwarfs that are really close to another star don't make great companions. This one used its immense gravity to literally suck the actual substance of its companion star onto itself. Sounds like bad vampire behaviour – sucking the very essence of life from its friend.

It dragged a stream of regular star stuff onto itself for about 16,000 years. This infalling star stuff sped up the rotation of the white dwarf until it was spinning at once every 200 seconds.

Now for the other dance partner. It is a neutron star, but it had a pretty spectacular past as well. In its earlier days it was massive – maybe eight times the mass of our Sun – and had evolved into an enormous red giant (or super-giant) star. And I mean *enormous*. To get an idea of its size when it was a red giant, if it had taken over the position of our Sun in our Solar System, it would have swallowed up Earth as it expanded outwards.

When the red giant collapsed, it exploded into a supernova, and in that cataclysmic event, it ejected most of its matter into space.

All that remained was a tiny neutron star, 1.27 times heavier than our Sun but only about 20 kilometres across! And spinning very quickly. How quickly? About 2.5 times every second. It was a special type of neutron star called a pulsar, which emits a regular pulsating beam.

So let's picture our two bizarre cosmic dancers, surfing on the floor of Space-Time.

Imagine in your mind's eye the two dancing objects (a white dwarf and a pulsar), really close to each other, travelling at around

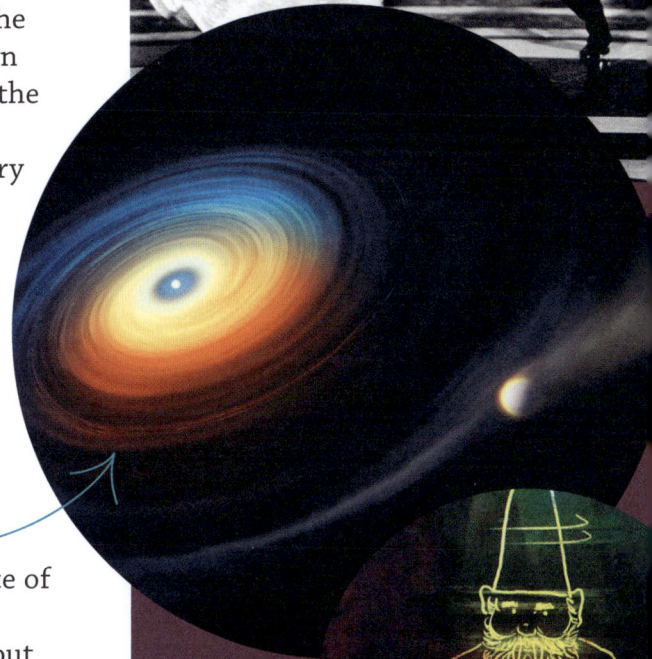

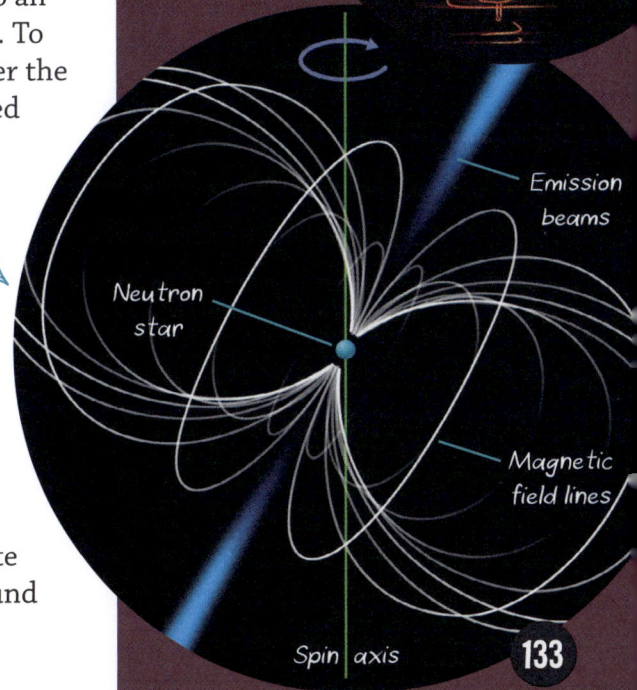

Emission beams

Neutron star

Magnetic field lines

Spin axis

133

a million kilometres per hour while orbiting each other every five hours, and each also spinning insanely quickly.

This is a very rare cosmic interplay, so astronomers have been studying these stars very closely for several decades.

FIRST, NEWTON ...

This dance involves truly extreme conditions. And that's just what you need for testing Einstein's General Theory of Relativity – which is a theory of gravity.

Way back in 1687, Isaac Newton published his Theory of Universal Gravitation. It was a magnificent achievement – brilliant, easy-to-understand and very elegant. It seemed to explain the motions of all the planets and moons in our Solar System. But after a few centuries, it became increasingly obvious that his theory could not explain what was happening with Mercury.

Mercury, like all the other planets, does not have a circular orbit. Instead, it has an elliptical or egg-shaped orbit – roughly similar in shape to a rugby or Aussie rules football, but not quite as squished. The point in its orbit where Mercury is closest to the Sun is called the 'perihelion'.

One slightly odd thing is that after each time Mercury orbits the Sun, the next perihelion shifts a little bit around the equator of the Sun.

Think of a daisy. Looking at it from above, it's got a central bit, with maybe a dozen petals coming out from it. (The daisy petals are way more elongated than the elliptical orbit of Mercury, but it's just an example.) Now let's pull off all the petals – except for one.

The elliptical orbit of Mercury around the Sun looks a bit like our daisy – with just one single petal. But with each orbit, that petal shifts around a little bit.

Astronomers have long known this. And they weren't surprised by it. Newton's theory of gravity could explain it very neatly – at first.

The physics and the mathematics tells us that if Mercury were the only planet orbiting the Sun, its perihelion would never shift. But there *are* other planets in the Solar System. So their masses combine to pull on Mercury with their gravity, and make the perihelion slowly track around the Sun. Technically, this is called 'precession', or advance, of Mercury's perihelion.

But by 1859, something very odd had shown up. Over time, the astronomers' observations about the Solar System had become more precise – observations such as masses, locations and orbits of other planets. And a little discrepancy popped up, which wouldn't go away. Newton's theory of gravity said the perihelion of Mercury should be over *here*, but it was actually over *there*.

The astronomers' telescopes would always see the perihelion happen 'over there' – a little off from where Newton said it should be, and always by the same tiny amount.

Astronomers tried all kinds of science to explain this. They looked for clouds of dust orbiting the Sun, or the gravity of another planet closer to the Sun than Mercury, the expansion of the Sun at its equator thanks to its spin, and many more desperate explanations.

But no matter what they did, the perihelion of Mercury was always a little bit away from where Newton's maths said it should be.

The shifting perihelion of Mercury as it orbits the Sun

THEN EINSTEIN 1 ...

But then, in 1915, Einstein published his General Theory of Relativity, and it explained the discrepancy. Perfectly.

Einstein had showed that Space-Time is curved. This meant that the distance around the Sun in curved Space-Time (which is what is really happening) is different to the distance around the Sun in flat space. This means that Mercury's orbit, which should be a perfect ellipse if it were following Newton's theory of gravity, was no longer an ellipse. The overall result of this was that Mercury's orbit traces out a pattern like a flower – with the perihelion shifting around with each orbit.

So Einstein won on this one, back in 1915.

THEN EINSTEIN 2 ...

So we are finished with the oddity in the orbit of Mercury, and are moving on to our white dwarf and neutron star in a crazy cosmic dance.

The movements in these dancing stars can be explained by a completely separate subtlety in Einstein's General Theory of Relativity. (Yes, Einstein was very deep.)

If the mass is spinning, the Space-Time gets distorted. This was shown by the New Zealand mathematician Roy Kerr, back in the 1960s.

Let me introduce you to Frame Dragging. 'Frame' means the

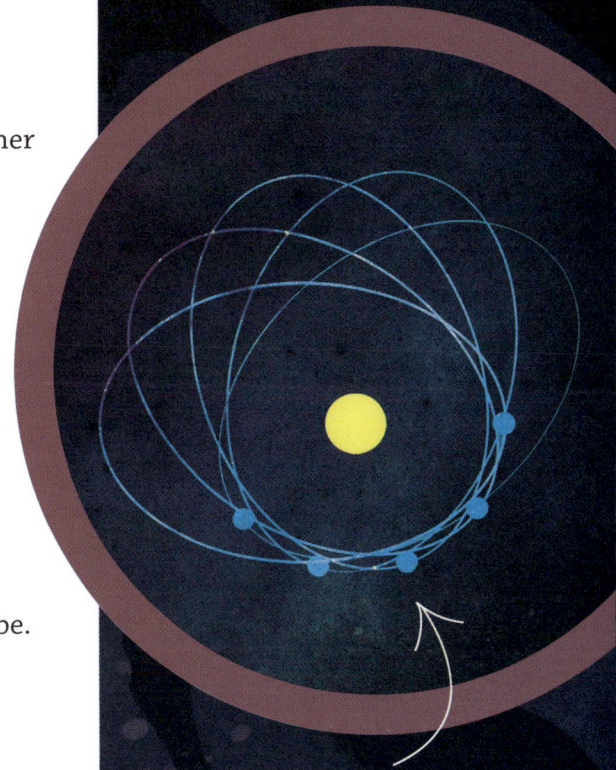

Surf faster than light

Lovers of science fiction (like me) get annoyed that it takes sooooo long to get anywhere in space – because nothing can travel faster than the speed of light (blame Einstein).

At 300,000 km/sec, light travels pretty quickly. However, the physicist Miguel Alcubierre has worked out (theoretically) how to travel faster than light.

Of course it involves surfing.

First, use 'exotic' matter (too complicated to explain here, but it has negative mass) to create a warp bubble of Space-Time. Then, shrink the Space-Time between you and your destination (say, a star 100 light years away). Simultaneously, expand the Space-Time behind you. Bingo, without you experiencing any acceleration, your little warp bubble will 'surf' on the distorted Space-Time to your destination at much faster than the speed of light. Hoorah!

Problems? A few.

We don't know how to make exotic matter. And we need a lot of it for just one trip – maybe millions of times the mass of the Universe! And you may well destroy your destination by the shock wave of your bubble arriving.

But these are just teething problems …

frame – or fabric – of Space-Time of our Universe. 'Dragging' means that this Space-Time is being dragged, or moved in some way.

As I said right at the beginning, the fabric of Space-Time has four dimensions – backwards-forwards, up-down, left-right, and time, which ticks away usually at one second every second.

You want to know what Space-Time is actually made of? Forget it! Too hard for a popular science book. Let's chicken out and totally avoid that Difficult Question.

Instead, I'll ask you to simply think of the fabric of Space-Time as being clingy and viscous – like honey. If it's viscous, when you move very quickly through the honey-like Space-Time, you can 'drag' it with you – and make waves in it. And this can explain the motions of the dancing stars.

So let's look at the implication of Roy Kerr's mathematics – when spinning objects orbit each other. I'll start with a 'simple' example of Frame Dragging.

DROP PLANET ONTO SPINNING STAR

First, imagine a universe that has only one single object in it. This object is a massive star – say, ten times the mass of our Sun – and is spinning very quickly. And this high-speed spinning is important. Like all spinning stars, it has an equator, and two Poles – North and South.

Second, let's introduce another object – a planet the size of the Earth. And let us just suspend this planet some distance away from this spinning star, directly above the North Pole. By the way, this planet is not spinning.

Now let go of the planet. The gravity of the rapidly spinning massive star will make the planet fall directly down towards the big star's North Pole.

Now under Newton's Theory of Universal Gravitation, our hypothetical planet just falls, splat! (like an apple from a tree), onto the North Pole of the spinning star.

On the way down – and this is critical – our planet does not start to spin. Why should it? After all, under Newton's theory of gravity, there is no force to make it spin.

But things are very different under Einstein's theory of gravity, better known as his theory of General Relativity. The rapid spin of the massive star drags the 'Frame of Reference' – the fabric of Space-Time – with it. The spinning fabric of Space-Time then 'drags' on our falling planet – and this planet begins to spin well before impact!

FRAME DRAGGING

Now, we can get back to our white dwarf star orbiting its neutron star partner, some 15,000 light years away. But the actual motions are more complicated. Not only are our stars spinning really fast, and orbiting one another, the whole spinning/orbiting system is also rolling, or tumbling, over and over!

Newton would never have put up with that kind of unforced motion. It simply could not be explained by his theory from 1687. But you guessed it – this odd behaviour *is* predicted by the 20th-century work of Einstein and Kerr, culminating in Frame Dragging.

EINSTEIN IS SPOT ON AGAIN – AWESOME AND BUMMER!

The good news is this is exactly as Einstein predicted way back in 1915. The bad news is this is exactly as Einstein predicted way back in 1915.

Let me explain. You see, there are some major problems in both astronomy and sub-atomic physics these days.

At the Big End of the Universe, we are very confident that both Dark Matter and Dark Energy exist – but we have no idea what they each are. That's one problem.

Another problem is with the four 'fundamental forces' (or 'fundamental interactions') that control everything in our Universe. They are the Gravitational Force, the Electromagnetic Force, the Weak Nuclear Force, and the Strong Nuclear Force. We can't blend them into a single theory, even though we've been trying for the last century.

Yet another problem is that, down at the Small End of the Universe, we know that particles called 'neutrinos' do have mass. But the standard model of sub-atomic physics says they should not have any mass. These problems are just the tip of the iceberg.

The physicists would absolutely love to discover something that starts off not making sense, but then morphs into the pathway that fixes some of our problems with the current physics. And unfortunately, proving Einstein correct yet again does not give us the breakthrough into the New Physics that physicists so desperately want.

With luck, we'll soon have some total intellectual wipe-outs that will help us surf into the new wave that physics absolutely needs to progress to the next level.

CARBS & ULTRA-PROCESSED FOODS

English is a funny language. Many of its words have two different meanings – like 'diet'. It can describe 'everything you eat', or it can mean 'eating to lose weight or eating/avoiding certain foods for medical reasons'.

'Sugar' is another one. To the general public, it's the granular white or brown stuff you spoon into your tea or coffee; but to biochemists, it's a whole bunch of related molecules with a sweet taste.

But 'carbs' (or carbohydrates) is not a word with two meanings. It's one of the three major macronutrients that food is made from, along with fat and protein. Carbs are one of the most important sources of energy for our bodies.

So I really don't get why people are so hostile towards and definite about 'giving up carbs', when what they are meaning to do is something quite different.

Sure, tell me you don't eat simple carbs (e.g. white bread, pasta or table sugar), or that you don't eat processed carbs (e.g. potato chips). But if you're still eating fruit and vegetables, you are still eating carbs. (And in terms of making healthy choices, eating more fruit and veg is on the money.)

Food is not all 'good' or 'bad' – after all, food does not have any intrinsic 'moral' value. But according to the World Health Organization, over the four decades between 1975 and 2016, the global prevalence of obesity nearly tripled.

There's lots of research and theories to try to explain this. Some of them are rather simple – and often, they are contradictory. And many of them tend to demonise, or sanctify, one particular food type.

Carbohydrates seem to confuse people. So let's start with the basics.

CARBOHYDRATES 101

Some people think that bread has 'carbs' but that fruit and vegetables don't. Let's run through some facts, beginning with small carbs and working our way through to big carbs in four steps – monosaccharides, disaccharides, oligosaccharides and polysaccharides. ('Mono' means 'one', 'di' means 'two', 'oligo' means 'a few' and 'poly' means a whole bunch.)

GREEN ROOM

Where have all the carbon atoms gone?

In case you are looking at the diagrams of molecules throughout this book and wondering where all the carbon atoms mentioned disappear to, here's a little secret.

Chemists adopt a convention when representing molecules in an illustration. It's related to the fact that we are carbon-based organisms. We carry lots of carbon atoms. Lots and lots. So if we had to write down every single carbon atom in a molecule, then diagrams of molecules would get very complicated very fast.

So here's the secret: to stop diagrams of molecules getting too messy, chemists just leave out the C representing a carbon atom at the points where two or more straight lines meet – they all 'know' that there is a carbon atom there. So when you see a drawing of a molecule, just imagine there is a carbon atom at these particular junctions. The carbon atoms haven't disappeared at all – they're there, but implied rather than explicitly shown.

MONOSACCHARIDES

The word 'carbohydrate' spells out what it is – literally 'carbo(n)-hydrate'. Let's start with a very small building block – a single carbon atom (C) joined to a single water molecule (H_2O).

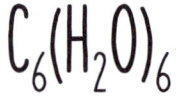

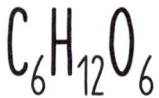

$$CH_2O$$

These very small molecules (in this case, a group of four atoms) don't usually exist by themselves. They normally join together, in groups of between three and eight.

Let's join six of these very small molecules together, to give us our basic monosaccharide – a pretty famous one, called glucose.

$$C_6(H_2O)_6$$

It's usually written as

$$C_6H_{12}O_6$$

This glucose molecule is shaped like a hexagon. In this view (the Haworth Projection), not all the hydrogen atoms are shown.

Other very common monosaccharides include fructose and galactose. There are many variations on this basic pattern of a monosaccharide. A few of them (such as deoxyribose) have slightly different numbers of carbon, hydrogen and oxygen atoms. Monosaccharides are also called 'simple sugars'.

Monosaccharides are small enough to cross the gut wall into the blood supply. They then end up in the liver to be further broken down, creating energy that can be used by our bodies. But some of them (especially glucose molecules) can have other destinations (such as the muscles) and other destinies (they can get stuck together loosely to make a small emergency supply of glucose molecules, such as 'glycogen', in case you miss out on dinner, or need extra energy while running a marathon).

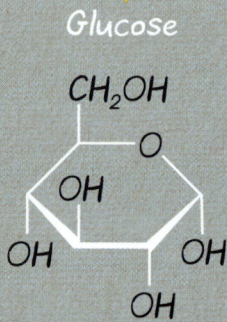

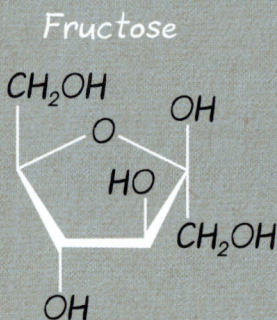

Glucose

$$CH_2OH$$

$$OH$$

$$OH \qquad OH$$

$$OH$$

Fructose

$$CH_2OH \qquad OH$$

$$O$$

$$HO$$

$$CH_2OH$$

$$OH$$

Psst! See the previous page – 'Where have all the carbon atoms gone?' – to find out the secret convention used by chemists in diagrams of molecules.

DISACCHARIDES

A disaccharide is two monosaccharides joined together. The process of joining two monosaccharides together means that you lose a water molecule.

There are lots of disaccharides in fruit and vegetables. One commonly consumed disaccharide is sucrose. Sucrose is made up of two monosaccharides – glucose and fructose.

Here's where the confusion happens for the general public. To a chemist, the word 'sugar' refers to any carbohydrate – from a monosaccharide to a polysaccharide. But for everyone else, the word 'sugar' refers to sucrose, the white crystalline water-soluble solid that we add to cakes and our cuppa.

The disaccharide in milk is lactose. It is made up of the monosaccharides glucose and galactose joined together. It's about one third as sweet as sucrose.

The disaccharide in germinating seeds is maltose (the name for two glucose molecules bonded together). It's also about one third as sweet as sucrose.

In general, disaccharides can be broken down in the small intestine into two monosaccharides, each of which again can end up in the liver, providing energy. (Or they can go directly to other tissues, such as the muscles.)

OLIGOSACCHARIDES

An oligosaccharide is just a slightly bigger bunch of monosaccharides joined together – between three and ten, to be specific. You're getting the idea by now. You can just keep joining those small monosaccharide molecules together to make bigger molecules.

Oligosaccharides include fructo-oligosaccharides (commonly called FOS), which are chains of identical fructose monosaccharides joined together. Many vegetables contain them. These oligosaccharides cannot be broken down in the small intestine, so they can't get to the liver as quickly as disaccharides do.

However, they can often be broken down in the large intestine, which means we can eventually get some nutritional benefits from them.

In the human body, oligosaccharides are more than just 'food'. They are often joined to proteins and fats. Because of this, they can have many other functions – such as in cell-surface receptors, cell binding and adhesion, immune response, and other roles.

POLYSACCHARIDES

Polysaccharides are a lot bigger. They are made up of thousands, sometimes tens of thousands, of monosaccharides. These can be joined together in long chains, or branches like a tree, or a combination of chains and branches.

Glycogen is a polysaccharide that combines both long chains and branches. It contains up to tens of thousands of glucose

This is the ball-and-stick model of a molecule of common house sugar. It has two monosaccharides joined together – glucose on the left, and fructose on the right.

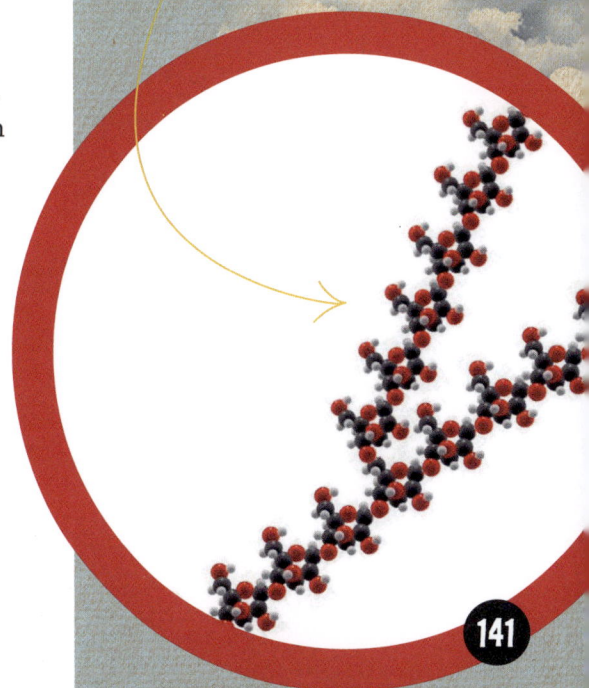

This shows the atomic structure of a single branched strand of glucose molecules, stuck together to make a much bigger glycogen molecule.

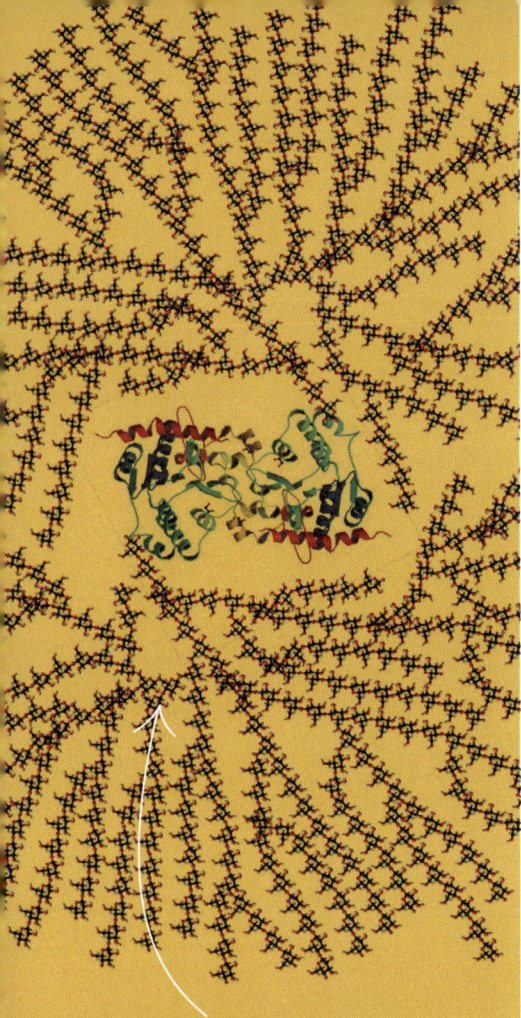

Schematic 2-D cross-sectional view of glycogen. Glucose molecules are arranged into many branches, which surround a core protein of glycogenin. The entire globular granule may contain approximately 30,000 glucose units.

monosaccharides joined together. Animals (including us humans) use glycogen as our main short-term energy reserve, and it's the main form of carbohydrate stored in the body. It can make up as much as 8% of the mass of the liver (immediately after a meal) and 1–2% of the mass of muscles. (However, fat is the major long-term energy reserve in the body.)

In general, polysaccharides are used in Nature as structural building blocks (such as cellulose or chitin) or to store energy (such as glycogen and starch).

Cellulose is built from thousands of glucose monosaccharides stuck together. But the glucose molecules are arranged in a different way from glycogen, to make a molecule with properties very much different from glycogen. Cellulose is often claimed to be the most abundant molecule in Nature, and is the prime ingredient of paper and cotton. Our human gut is unable to break down cellulose into its component glucose monosaccharides.

However, termites can digest cellulose. This is not because the termite gut can inherently digest cellulose by itself, but because all termites have been colonised by various microbes in their gut – such as symbiotic protozoa, flagellate protists, spirochetes, and many other microbes not found in any other known creatures. It's these microbes that break down the cellulose into its individual monosaccharides. The microbes get a meal, and they share some with their host, the termite. Starch is yet another polysaccharide – again made of thousands of glucose molecules, but arranged in a different way from glycogen and cellulose, and joined together with different chemical bonds (such as glycosidic bonds). It is used by plants to store energy. Potatoes are mostly starch. We can't break down this starch until the potato has been cooked, weakening the bonds joining the glucose molecules together.

Carbs are interesting, and complicated. But they are also an essential part of our diet – which means we can't fairly blame carbs for worldwide obesity.

So how about an alternative theory? Could part of the problem be the over-processing of the food we eat?

PROCESSING 101

I'll start with an overview.

First, there are 'unprocessed' foods. These include any of the edible parts of plants – such as the roots, the leaves, the fruit or the

seeds. They also include animal meat, dairy and eggs. Sometimes, these foods have undergone a tiny amount of 'processing' – a bit of pasteurising, freezing or drying. But they are still counted as unprocessed foods, because they come to you without any added fats, salts or sugars.

Second, there are 'processed' foods. These do come with fats, salts or sugars added – sometimes to increase their shelf life, and sometimes to sharpen their flavour. These foods include meats that have been cured and/or salted, fish or vegetables that have been canned, as well as cheese and fermented drinks, such as beer or wine.

Third, there are 'ultra-processed' foods. Hang on, doesn't 'ultra' mean 'super good'? That depends on where you sit – either in the marketing camp or the nutrition camp. You can usually eat these straight out of the packet or tin, without any cooking. This is because they typically have a long shelf-life – especially when compared with fresh foods.

They have lots of additives, including oils and solid fats. But that ain't all – there are chemicals to enhance colour and flavour, other chemicals to bulk out or firm the foods, sweeteners (sugar and non-sugar), and so many more.

The food industry deliberately aims to make 'ultra-processed' foods 'hyperpalatable'. They want you to crave the food, but to never get full, so that you keep eating (and buying) more of it. How sneaky of them – especially when these foods tend to be bad for your long-term health.

BLISS POINT

The word 'hyperpalatable' means that you keep on wanting more. These 'artificial' foods have been manufactured (that's the correct word) to taste so much more tempting than foods made by boring old Mother Nature. It takes a lot of clever chemistry to do this. The food chemists talk about the 'bliss point' in the flavour of a food.

For example, if you add sugar to water, at a certain percentage, the sugared water tastes wonderful. Weirdly, however, if you add a little more sugar beyond this point, the taste gets worse.

Now, here's where the food chemists step in to do their stuff.

They keep the added sugar but also toss in salt, fat and a dash of bitter. Amazingly, the sugared water tastes even better than before. With a bit of juggling, the food chemists can reach the 'bliss point' – the maximum amount of taste and flavour. The hyper-

143

taste is 'better' than the original food – and as a bonus, the extra ingredients are cheap, which keeps costs down.

In a nutshell, the big food companies make their 'ultra-processed' foods by breaking food down into its basic chemical ingredients – and then reconstituting them into foods that never existed before.

For example, natural foods very rarely have both fat and carbohydrate together in large quantities – it's usually one or the other. And carbohydrate in Nature nearly always comes packaged with lots of fibre.

But the food chemists don't respect this natural balance, which took millions of years of evolution. They do their divide-and-conquer stuff to create very unbalanced foods.

POTATO TO SUPERPOTATO

Look what ultra-processing can do for the humble potato. The food chemists call it 'loading and layering'.

A potato is a bunch of monosaccharides joined together. Food chemists will *load* the potato with fat (which is very cheap to buy). To do this, they cut the potato into thin chips, giving them a huge surface area, to better absorb the fat. They deep-fry the chips, so the fat gets loaded into the very 'flesh' of the potato – in between the carbohydrate molecules. Sometimes they even double-fry the potato.

Then they *layer* the deep-fried potato chip with cheese – to add more fat. Shame about the extra calories/kilojoules.

They finish it off by adding buckets of salt, artificial flavours and maybe some sugars or sour cream.

Suddenly, the humble potato has turned into a super-delicious item that is loaded with extra calories/kilojoules – but not much extra nutritional goodness.

For the food companies, the lovely thing about these hyperpalatable, ultra-processed foods is that *they are their own reward*.

You'll keep on eating, even after you're not hungry anymore. That's because ultra-processed foods are linked to emotional pleasure – not hunger. And best of all, they don't make you feel full, so you'll eat even more.

LINKS TO OBESITY?

Not surprisingly, these flavour food bombs are popular.

Back in the 1970s, the big food companies in the USA began manufacturing 'hyperpalatable' foods. That's also when the obesity epidemic began. Maybe it's a coincidence, or maybe it's not?

In the USA today, about 60% of all the energy in the food that people eat comes from these ultra-processed foods. That is an astonishingly high percentage. These foods include energy bars, sugary breakfast cereals, chips, pastries, fruit-flavoured and fizzy drinks, and so on. The Australian figures are 42% for ultra-processed foods, 35% for unprocessed or minimally processed foods, and 16% for processed foods. Various 'culinary ingredients' (sugar, plant oils, butter, etc) make up the rest.

Other studies showed that people who consume these ultra-processed foods eat about 25–40% more calories or kilojoules each day than people who eat unprocessed foods. That's mainly because we don't really know what we are consuming when we eat these manufactured edible products. When we ingest them, there's a major disconnect between how much energy our senses think *should* be delivered to our gut, and how much is *actually* delivered to our gut. According to Dr Barry Popkin, a Professor of Nutrition at the University of North Carolina, 'ultra-processed foods are not only more seductive, but people eat more of them'.

Will our descendants be able to evolve so that their bodies can benefit from these substances that are sold as food? After all, given enough time, humans can adapt to the foods available to them.

The Inuit peoples of the Arctic do quite well on their high-fat/low-carb diet. The Japanese eat an opposite diet – low-fat/high-carb – and have long, healthy lives.

But perhaps the ultra-processed food diet is so far removed from natural foods that we humans won't be able to evolve and adapt to it, and still be healthy.

145

SPIDERS CAN DESIGN, BUILD —AND COUNT!

Spiders are more than just terrifying bundles of black hairy legs. They plan for the future, they make and use tools, and — amazingly — some of them can count.

And, unfortunately, most of us respond to spiders by trying to stomp on them.

SPIDEROLOGY 101

Spiders are very successful, even without our 'love'.

There are more than 48,000 different species of spiders, and, on average, about 130 spiders on each square metre of dry land. They are also among the oldest life forms on land — much older than the dinosaurs, which started up about 220 million years ago. Most of the major groups of animals began appearing in the fossil record during the 'Cambrian explosion', about 541–520 million years ago – they all lived in the oceans. Animals started appearing on land only about 484 million years ago, and terrestrial spiders followed later.

A big element of spiders' success is their ability to make silk, and they've been making it for around 350 million years. A spider's silk comes out of specific organs on the body called spinnerets.

We humans use about 20 common amino acids to make the proteins our bodies need, but spiders are much more economical in producing their silk. They mostly use just two amino acids (glycine and alanine) that are arranged in long 'chains', i.e. crystalline sections joined with 'amorphous linkages'.

This silk is a really good starting point to appreciate how smart a spider really is. Spiders don't just make one single type of silk. They tweak the biochemistry of the silk, depending on its intended use. Some spiders can make up to seven different types of silk. One type of silk can make a web hanging in your garden – lovely to look at, but a death trap for flying insects. Other types of silk spiders produce are used for climbing, for flying, for making protective cocoons for their eggs, as well as for building and lining the nest for some 'primitive' spiders, for immobilising prey,

Spinneret of Australian garden orb weaver spider

as a source of food, as alarm lines to let the spider know that a small meal is nearby for spiders that don't make actual full orb webs, and more.

About one third of spider species make the 'classic' circular orb webs. These spiders appeared relatively recently in the evolutionary record – only about 110 million years ago. Unfortunately for the hungry spiders, about half the prey that runs into an orb web will successfully escape.

PLANNING FOR THE FUTURE

Spiders can construct a web that is super efficient in three ways at the same time – filling the available space, avoiding obstacles and catching food. And the amazing bit is that spiders can do this even though most of them are almost completely blind.

Spiders' webs need to be as efficient as possible, because spiders do not go out and hunt for their food. Instead, they use a sit-and-wait strategy.

Many spiders start their web by throwing out a horizontal strand of silk across a gap between two objects, such as trees or the walls of your house. Then they map out the space for their web, by descending from and ascending to many locations anchored by this first horizontal strand. This 'mapping' is surprising because, until recently, we thought that only mammals and birds could mentally represent a space, and make what is technically called a 'cognitive map'.

Turns out, our arachnid friends are in the Elite Mental Map-Making Club as well.

It really looks like spiders can actually think ahead. They will alter the size and shape of their web, based on past experience, to suit future circumstances. They will lay down less silk if they are running low on reserves of silk, or if the external temperature drops and so they are running the risk of dying from cold. Or, if some specific strands caught lots of insects last time, they will adjust the tension and stickiness of the silk in that section of the web, to further improve the web's catching ability. They will even lay down extra strands if something big broke through the web last time, or if the local prey are larger than they used to be.

So their webs are not built automatically, by primitive, innate

POINT BREAK

STRONGER THAN STEEL

How does a spider produce a material, silk, that is stronger than steel – both in comparisons of weight-for-weight and volume-for-volume? And how does it make this silk at room temperature, with common biological chemicals?

We still don't fully understand. But we do know that the 'science magic' happens in the spider's internal silk glands and external spinnerets. Most spiders have six spinnerets, usually on the underneath of the spider's abdomen, at its rear end. (But some spiders have two, four or eight spinnerets.)

The spinnerets can work and move together, or operate independently. A single spinneret has many microscopic pipes (or spigots), each one of which manufactures a single filament. These single filaments can then be blended or woven in numerous ways to make a single thread, which allows the spider to spin a variety of silks for many different purposes.

A large part of the spider silk's strength comes from how the various proteins in each filament are manufactured and lined up in very specific orientations.

Unusually long spinnerets

No spiders = no agriculture

Like me, you (and some 3–6% of the population) might be irrationally scared of spiders. The psychologists still haven't given us good explanations for what causes arachnophobia (an irrational fear of spiders). It might be from our past evolutionary history, or from our current culture – we just don't know where this fear comes from.

But remember this. If it weren't for our little friends with eight legs, we would have no agriculture. According to the evolutionary biologist Miquel Arnedo, from the University of Barcelona, without spiders 'you couldn't have any crops – insects would eat them all'.

Spiders do a lot of free biological control for us. They eat about 400–800 million tonnes of insects each year, worldwide.

Spiders need to eat, and we humans accidentally benefit from this.

149

reflexes, in the same way every time. Instead, spiders continually change and improve their webs.

This is amazing behaviour, especially when you consider how small the brain of some spiders is. Some orb weaver spiders weigh less than one thousandth of a gram, and the brain is a tiny fraction of that.

SPIDERS CAN COUNT – PAST TWO

One jumping spider in the genus *Portia* can do amazing stuff with its brain: it can remember, plan and count!

These spiders attack other spiders up to twice their size. Tiny predators (they are 5–10 mm long), they silently roam the landscape, looking for something to eat – so, unlike other spiders, they have great eyesight.

In one cleverly designed study, scientists put a single *Portia* spider on a tower on a tiny island in a tiny lake (okay, it was just a puddle!), surrounded by dry land. The distance between the tower and the dry land was too great for them to bound across in a single leap.

Now, *Portia* spiders hate getting wet. So the experimenters put two floating pathways across the lake. And outside the lake, on dry land, they put food – another species of spider – for the *Portia* to eat, on another tower. But (and here's some of the clever design of the experiment) when the *Portia* went down its tower, it lost sight of its potential dinner. So it had to use its memory.

THREE EXPERIMENTS

First experiment: memory.

One floating pathway led to dinner, while the other didn't. The *Portia* could see both pathways from its tower. And after it had climbed down from its tower, it always remembered which was the correct pathway that led to dinner. Yup, the *Portia* remembered its mind map.

Second experiment: memory and efficiency.

Both floating pathways led to dinner, but one was shorter. Yup, the *Portia* chose the shorter pathway. Again, the *Portia* could remember the mind map, so it could reach its dinner more quickly.

Third experiment: counting.

For a bit of background, you need to realise something well known to psychologists – a sudden pause in a creature's activity indicates it might be 'confused' or 'surprised'. Using this knowledge, if you design an experiment cleverly, you can work out (for example) what a baby is thinking.

So, getting back to our *Portia* spider. Up on its tower, it sees a few spiders, just hanging around waiting to be eaten, on the other tower. So it abseils down from its tower, crosses the water and arrives at dinner. But while the *Portia* is in transit, the cunning experimenters change the number of spiders waiting to be eaten. There might be more, or there might be fewer – but the point is, the number of spiders for dinner is different.

And get this – the *Portia* spider suddenly pauses!

This means that it's surprised. It expected a certain number of dinner items, but it didn't find them.

Yes, this really suggests that the *Portia* spider can count. It doesn't count to very high numbers. *Portia* seems to have just three 'numbers' – one, two and 'more than two'. Even so, it *is* counting.

(Of course, once the confusion/surprise is over, and the *Portia* accepts the mathematical paradox, it just starts attacking and eating. After all, why waste a good feed, simply because of a little theoretical mathematics?)

Portia spiders can count their dinners, leaving us to count our blessings as they protect our crops.

So if you ever see a spider and have a reflex urge to thwack it – try to resist. As well as protecting our crops and being an artisanal builder, it probably knows where you live …

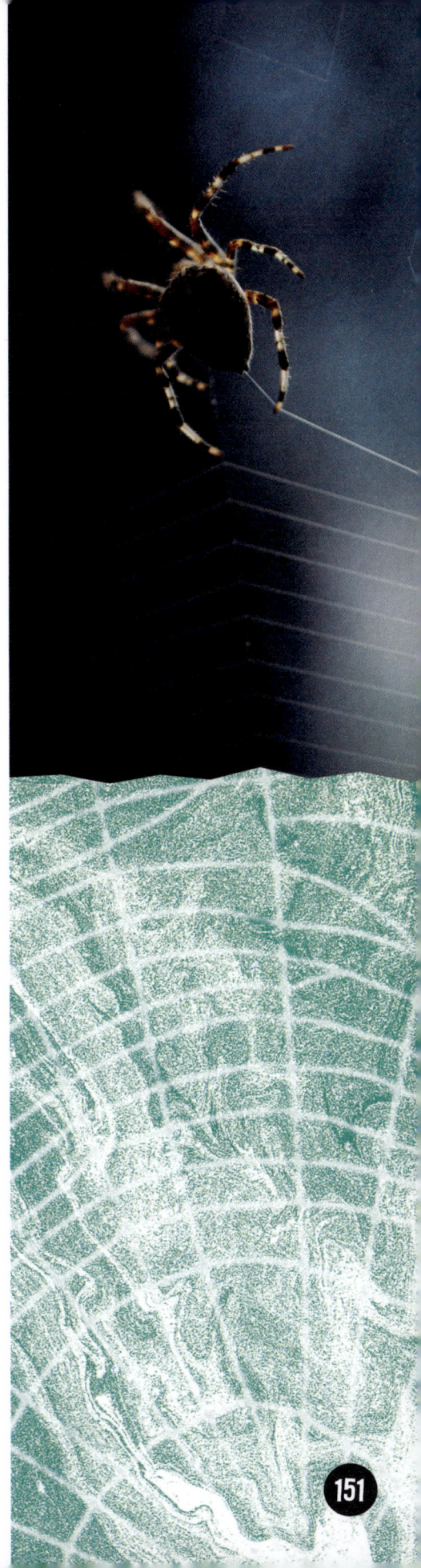

WINDED –
A BREATHTAKING
EXPERIENCE

I remember the very first time I got winded, because it really frightened me. I was about nine years old, and I was running with my dog. Suddenly, with absolutely no warning whatsoever, my dog ran in front of me and tripped me over.

I fell forward, flat, onto the long grass. Luckily, I didn't break any bones, or even get any bruises.

But as I lay facedown on the soft grass, I suddenly realised that I could not breathe!

GASPING FOR AIR

For the next few seconds, I was mostly confused and scared.

Intellectually, I knew that I had been running and burning up lots of energy, so I needed to breathe, but for some reason I physically *could not breathe*! My intellectual analysis was overrun, replaced by pure panic. I really, *really* needed to breathe, but I couldn't – so I thought I must be going to die.

After what seemed an eternity, but was almost certainly just 15 or 20 seconds, I was able to take a strangled breath. I lay there, gasping quickly on the ground, and after a minute or so, I was able to roll over onto my back and, a little later, actually sit up.

Gradually, my breathing slowed down and returned to normal.

So what had happened?

Well, I had been winded. The name 'winded' makes it sound like all your wind has been pushed out of you.

But has it? Nope.

So what's going on? Let's start with a little bit of anatomy and physiology.

RESPIRATORY PHYSIOLOGY 101

You have two major cavities, or hollow places, in your trunk.

The upper cavity is called the thoracic or chest cavity, and it's surrounded by your rib cage. It's mostly filled by your lungs. (As an aside, there's another much smaller cavity inside your chest

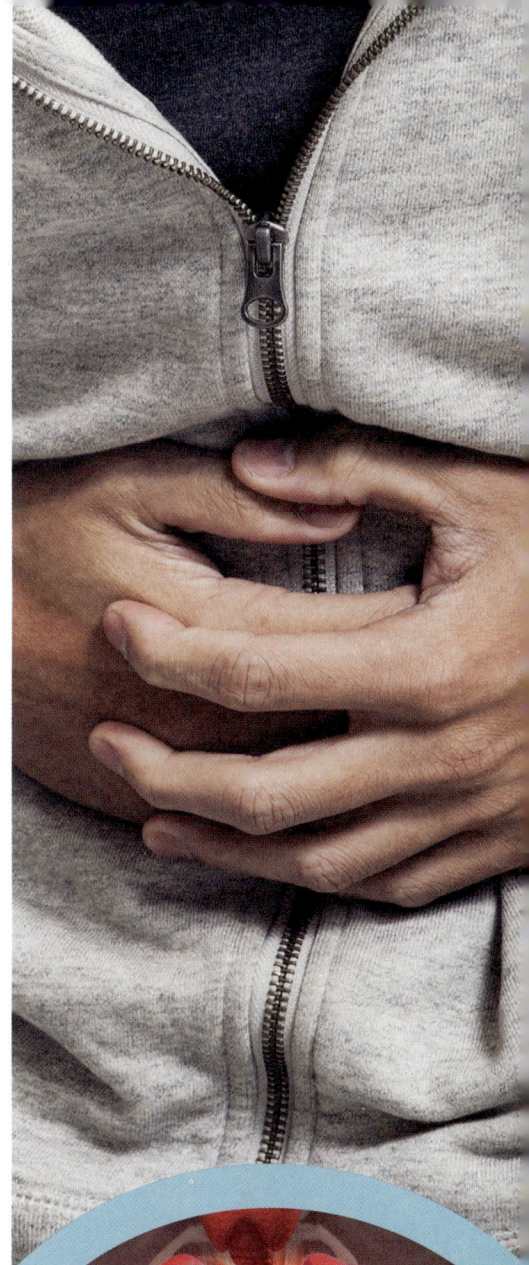

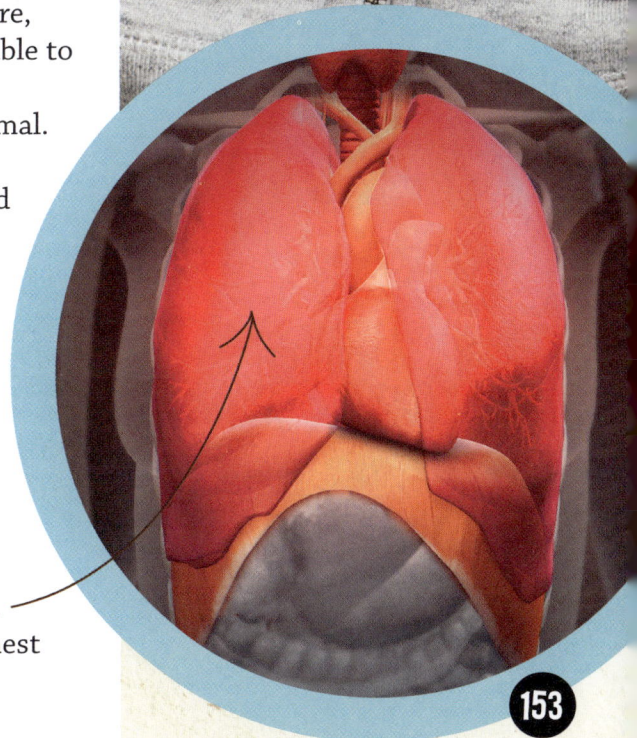

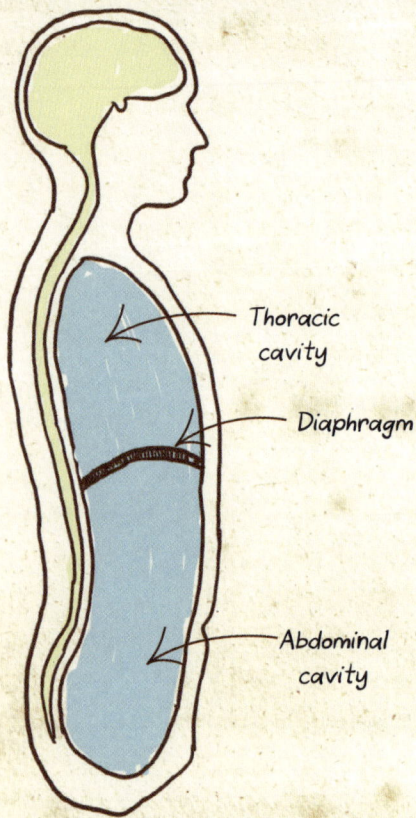

cavity. It's for the heart, but it almost certainly has nothing to do with being winded – so let's just ignore it.)

The other major cavity starts below your lungs, and is called the abdominal cavity. The abdominal cavity runs from the bottom of your ribs down to the top of your legs. And it's chock-full of goodies like the 10 metres of your gut, and organs such as the liver, spleen and pancreas, as well as various reproductive organs.

Now between these two cavities is an unusual muscle called the 'diaphragm'. When you get winded, it's probably because the diaphragm has stopped working normally.

The diaphragm is different from most other muscles that you have some control over – the so-called voluntary muscles. Most of the voluntary muscles in your body are shaped like long cords or strings – but the diaphragm is like a thin sheet and curved into a dome shape. The diaphragm runs across the bottom of your rib cage, from your chest to your lower back.

Your diaphragm can be consciously controlled by you (e.g. if you hold your breath), but most of the time it's run by the respiratory centre in your brainstem (at the top of your spinal cord).

When your respiratory centre sends the right kind of electrical signal to your diaphragm, this muscle contracts. It changes shape, from being curved like a dome to more flat. And remember that located directly above the diaphragm are the lungs, so the bottom section of the lungs expands downwards as the diaphragm flattens.

This expansion creates a lower pressure – a partial vacuum – inside the lungs. So air then rushes through your mouth or nose, down your windpipe into your lungs – and, behold, you have just taken a breath!

A few seconds later, the electrical signal from your respiratory centre switches off, and the diaphragm relaxes back into its natural dome-like shape. The lungs shrink, and some of the air inside your lungs is pushed back up your windpipe and out through your mouth or nose.

So now you've breathed in some air, and then you've breathed it out again – you've had one complete breath.

The breathing cycle repeats around 15 times each minute. And during that minute, you move about 5 litres of air into, and then out of, your lungs. This air is sucked in and then pushed out by the rhythm of your diaphragm contracting and relaxing.

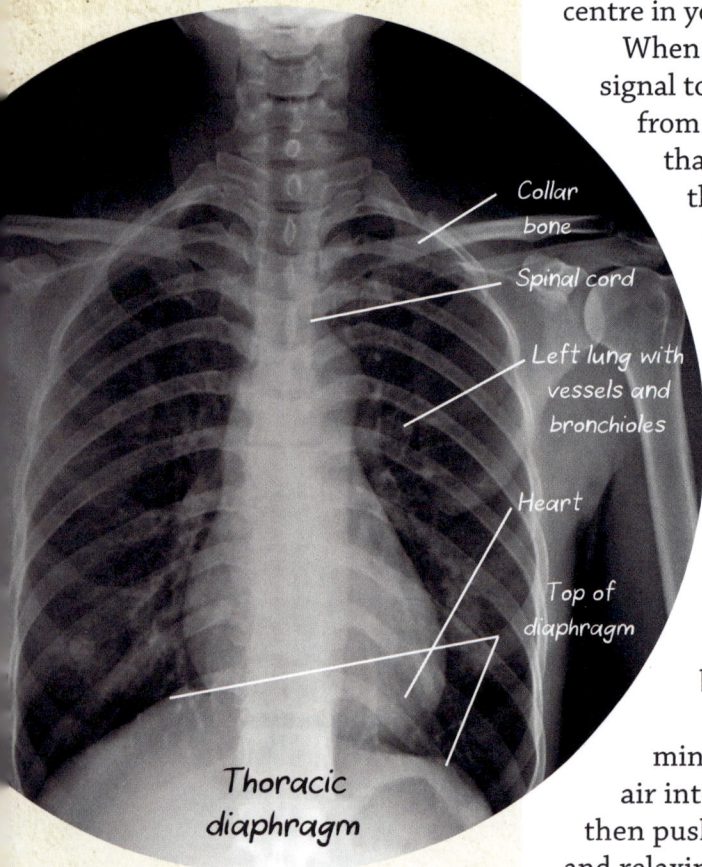

Thoracic cavity

Diaphragm

Abdominal cavity

Collar bone

Spinal cord

Left lung with vessels and bronchioles

Heart

Top of diaphragm

Thoracic diaphragm

WINDED – CRAMP?

So what's going on when you are winded?

If you get some kind of blow to the chest or the abdomen – in the case of my nine-year-old self, falling flat onto the grass – you might be winded.

The blow generates a sudden pulse of high pressure inside your chest cavity, and/or your abdominal cavity. It seems that this spike of high pressure sends your diaphragm muscle into a spasm. It can't contract any further, and it can't relax from where it is. It's paralysed.

It's a bit like getting a cramp in a leg muscle. With a leg cramp, for a short but painful time, you can't straighten or bend your leg – it's 'frozen' in place. In the same way, when you get a spasm in your diaphragm, you can't contract it or relax it – it's 'frozen' in place. So you can't breathe in or out for a few seconds. But after a short while, just like a leg cramp, the spasm goes away.

Now, we're not exactly sure why the sudden pulse of high pressure can cause a spasm in the diaphragm. After all, you don't get winded every time you fall flat onto your chest. Maybe the spasm is caused by the spike of high pressure being focused on a sensitive part of the diaphragm? Or maybe the high pressure acts on some nerves in the abdomen (the solar plexus), and this somehow causes the spasm? We simply don't know.

Impact causes diaphram to spasm, preventing lungs from filling with air

TREATMENT

If you do get winded, you might find some relief from sitting up and leaning forward, and then using your abdominal muscles to gently force your belly out and back in. When you do this, your diaphragm muscle flattens and morphs back into its curved shape – and the lungs will follow, shifting air into and then out of your lungs. It won't be as good as your diaphragm actively contracting and relaxing, but at least you'll be moving some air in and out of your lungs – and doing something active will help stop you from panicking.

And don't worry if you forget to do this. The spasm will go away all by itself, and your diaphragm will be back and shifting normally in no time at all.

In the meantime, it won't help to meditate – this is probably not the right time to follow the mindfulness mantra of 'focus on your breath'. Forget about breath – just force your belly in and out, and before you know it, your spontaneous breathing will follow.

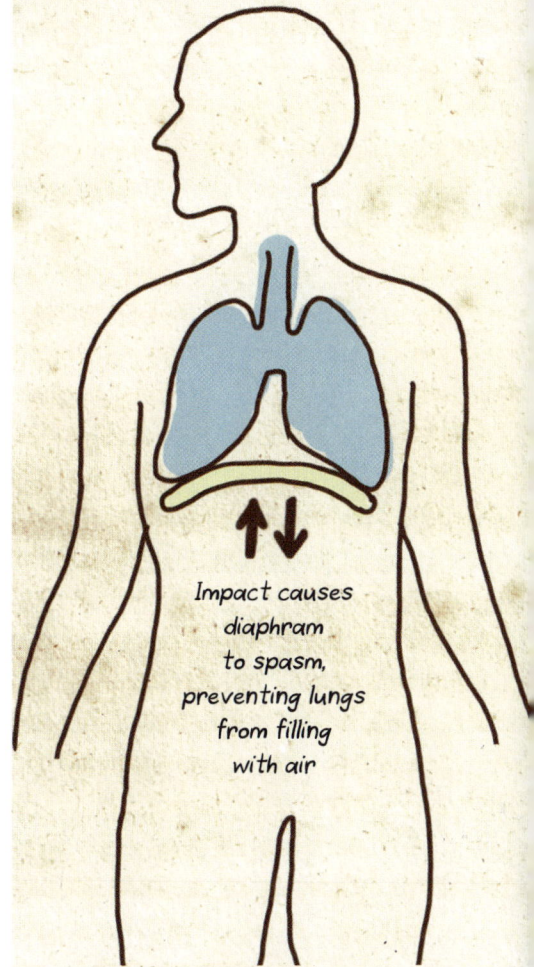

155

MIGRATING PLANETS — A DIFFERENT TACK

Humans have been looking at the stars for hundreds of thousands of years. For most of that time, we thought we were the centre of the Universe. Only a few hundred years ago did our scientists work out that the Earth orbited the Sun – but back then, we still had no idea other solar systems existed.

However, we now know of over 4,000 planets orbiting other stars in more than 3,000 solar systems. We've found so many that we even had to invent another name for them – Exoplanetary Systems.

Naturally, finding all these newbies led to seeing how they compared with us. They turned out to be different in lots of ways.

So then came the big and obvious question, 'How come our Solar System is so different from practically all the Exoplanetary Systems we've found out there?'

WHAT MAKES US DIFFERENT?

Yes, our Solar System is very different from Exoplanetary Systems.

Exoplanetary Systems mostly have planets that are all fairly big and that nearly always orbit very close to their host star.

But our Solar System has planets of many different sizes scattered evenly from Mercury close in to our Sun to Neptune, way out in the sticks.

Why the different layout?

The simple answer is that we don't know yet, though we do have some theories.

A CLOSER LOOK

Let's look more closely at how different our home Solar System really is from Exoplanetary Systems. As a reminder, our arrangement is four smallish rocky planets close to the Sun (Mercury, Venus, Earth and Mars), and the four much bigger planets further out. Jupiter (318 Earth Masses) and Saturn (95 Earth Masses) are truly huge, and more than 90% hydrogen and

DR KARL'S Q+A

What are shooting stars?

Shooting stars are not stars – but they are as bright as stars, and they shoot across the sky. Some of them are little rocks left over from the collisions that happened way back in the early days of our Solar System, some 4.7 billion years ago.

By the way, when they are in space they are called 'meteoroids', when they travel through our atmosphere they are called 'meteors', and when you find one as a little rock on the ground it's called a 'meteorite'.

They run into our upper atmosphere at speeds of up to about 30 km/sec. They are moving faster than the air molecules can get out of the way (about 300 m/sec), so a shock wave happens. The pressure in front of the meteor builds up, and both the air and the meteor begin to get very hot. (Think of pumping up a bike tyre with a hand pump. Pretty soon, the pump gets hot.) Usually the meteor begins to light up when it's about 90 km up.

Each year, about 20,000–50,000 tonnes of meteorites land on our planet.

157

TO MEASURE IS TO KNOW

Lord Kelvin, a brilliant physicist, famously said, 'To measure is to know.'

He also spelled it out with a longer version, 'When you can measure what you are speaking about, and express it in numbers, you know something about it; but when you cannot measure it, when you cannot express it in numbers, your knowledge is of a meagre and unsatisfactory kind.'

William Thomson (aka Lord Kelvin) was one of the most important physicists of the 19th century, as well as a successful inventor and wealthy businessman.

At the University of Glasgow, he was its youngest ever student (at age ten) and, later, also its oldest ever student, re-enrolling immediately after he retired at the age of 75.

He was the first to build a machine to quickly and easily measure the depth of the ocean. The machine used piano wire wound onto a drum, and a very heavy lead weight. His machine made possible the successful installation of the first transatlantic telegraph cable between Ireland and Newfoundland in 1858–66.

Another example of his being in the lead of technology was that in 1881, his home in Glasgow became the first house in the entire world to be fully illuminated only by electricity – 106 light bulbs. (Though, historians disagree a little with each other on these points – the first or *one* of the first, or 106 or 112 lamps? Ah well …)

He was part of the movement in physics that led to the very deep understanding that 'heat' is the 'energy equivalent of work'. In 1848 he coined the word 'thermodynamics'. In fact, the Kelvin Temperature Scale is named after him.

Part of measuring is using the right units. When we think of distances to the stars, we think big and use light years, not centimetres. Inside our Solar System, astronomers use 'the distance from the Sun to the Earth' as a unit of measurement.

He was also one of the first ever scientists to be ennobled in the United Kingdom, receiving the rank of Lord. In 1866 he became Sir William Thomson and, later, Lord Kelvin. He chose the name Kelvin from the River Kelvin that winds around part of the University of Glasgow campus.

helium – so they're classified as gas giants; Uranus (14.5 Earth Masses) and Neptune (17 Earth Masses) are still big, but contain less than 20% hydrogen and helium, and so are classified as giant ice planets.

In most cases, Exoplanetary Systems fall into one of two patterns: either they have close-to-the-star super-Earths or they have a close-to-the-star Hot Jupiter. Super-Earths are planets more massive than Earth but less than Uranus and Neptune. (For various reasons, planets that are close to their host star are usually easier to find than planets that are a long way out.)

How close is 'close'? Let me introduce you to the Astronomical Unit, or AU. We call the distance between the Earth and our Sun one Astronomical Unit, or 1 AU (it's just under 150 million km). Mercury is about 0.4 AU from our Sun, Venus is 0.7 AU, Mars is 1.5 AU – you get the idea. I'll mention Jupiter (at 5.2 AU) and Saturn (9.5 AU), because they are major players in our story.

Now that we have that out of the way, let's look at the three major differences between 'our Solar System' and 'Exoplanetary Systems'.

First, our Solar System doesn't have any super-Earths. But most of the Exoplanetary Systems we've found *do* have a bunch of super-Earths, usually very close in to their parent star. Kepler-11 is a star almost identical to our Sun, and is about 2,150 light years from us. It has six super-Earths, with individual masses ranging between 1.9 and 25 times the mass of the Earth – with a combined mass of about 47 Earths. (Yes, it is a bit of a stretch to call the planet that is around 25 Earth Masses a super-Earth, but we are still in the early days of Exoplanetary System research.) These six (or five) super-Earths all orbit Kepler-11 at distances between 0.1 AU and about 0.5 AU. In fact, five of them are closer to their parent star than Mercury is to our Sun.

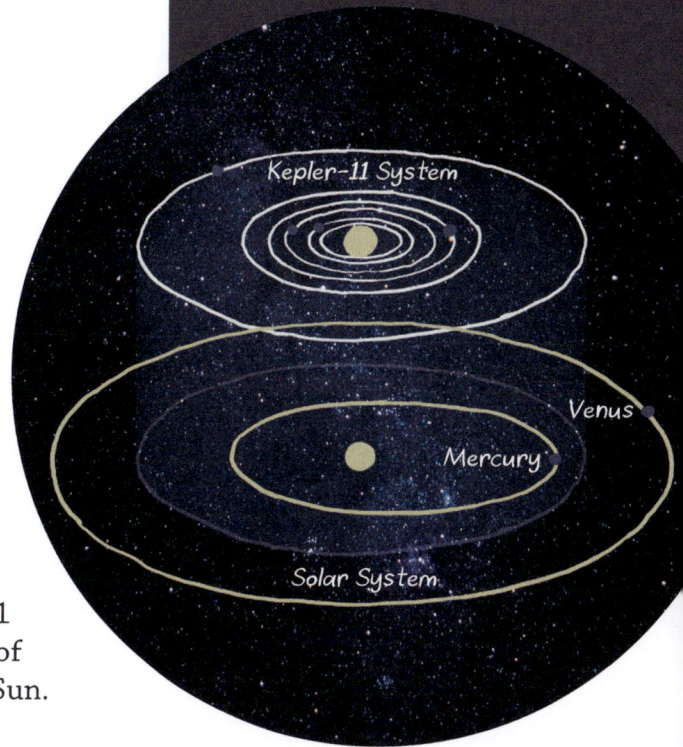

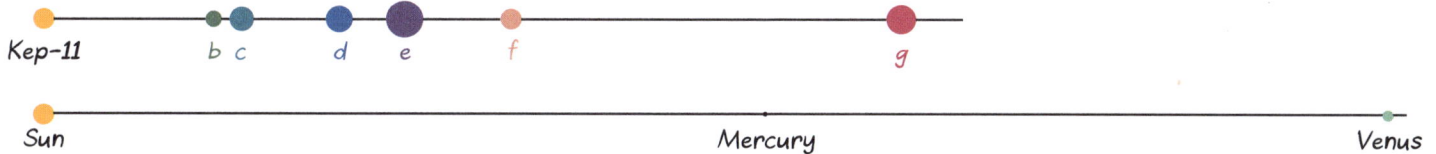

Second, in Exoplanetary Systems that don't have super-Earths, we sometimes find a single big planet, like our Jupiter. The difference is that while our Jupiter is a long way from the Sun (5.2 AU) and takes 11.9 years to complete its orbit, these Exoplanetary Systems have their single Jupiter-sized planet really close in to their host star. These exo-Jupiters usually orbit their host star much closer than does our planet Mercury (our closest planet to the Sun, at 0.4 AU), which completes an orbit in 88 days. Some of these planets are close enough that they orbit in less than ten days. When they're this close to their host star, they're really hot – and get the fabulous name of Hot Jupiters! (Just as a reminder, our Jupiter is way too massive to classify as a super-Earth.)

Third, the overall layout is different. We usually find that Exoplanetary Systems mostly have their planet(s) nestling in close to their host star – usually closer than Earth or Mercury.

But our Solar System is the opposite. There are no planets closer than Mercury. Furthermore, we have planets all the way out to Neptune, about 30 AU out from the Sun.

So we have all these issues with our Solar System:
• zero super-Earths;
• zero Hot Jupiter(s);
• our Jupiter is a long way from its star;
• too many planets, coming in too many different sizes;
• too many planets, with very big orbits.
How come?

GRAND TACK – SHORT VERSION

While the simple answer is that we don't know, one interesting suggestion is called the Grand Tack. In this case, the word 'tack' comes from sailing, and means 'a change in direction'.

The Grand Tack hypothesis starts with Jupiter coalescing from the primeval gas and dust of the early Solar System, around 3–4 AU from the Sun (somewhere between the present positions of Mars and Jupiter). Then Jupiter, fully formed, migrated inwards towards the Sun to around 1.5 AU. A little later, Saturn followed it towards the Sun.

While they were close to the Sun, Jupiter (and probably Saturn) became the giant 'wrecking ball(s)' of the early Solar System – getting rid of the super-Earths in the process.

Jupiter then changed direction with its Grand Tack (like a sailboat changing direction) to migrate outwards to its present location of 5.2 AU.

But Jupiter absolutely needed the help of Saturn (more about this below) to wave goodbye to the Sun, and head back out into the outer Solar System. Otherwise, it would have remained close in to the Sun, all by itself – having destroyed the super-Earths.

GRAND TACK – LONG VERSION

Different versions of the Grand Tack hypothesis give different event timings. This is a middle-of-the-road 'beige' version.

Let's go back to when the Sun coalesced from a giant cloud of gas and dust, and started burning its nuclear fires.

Way back at the birth of the Solar System, as the Sun began to spin, another cloud of gas and dust flattened into a disc, also spinning, and also orbiting the Sun. The disc survived for a few million years. Along the way, Jupiter coalesced from this disc of gas and dust.

HOT JUPITER

Why the very rock-and-roll nickname, *Hot* Jupiter?

Because it's so close to its host star that its 'surface' temperature is really hot. The surface temperature runs from around 1,000°C – roughly the temperature of molten lava – to 4,000°C and more. This is a lot hotter than the surface temperature of our Jupiter, which is much colder, around -110°C.

Three points here about 'Jupiters' and 'hot'.

First, if a planet is mostly made of gas, as our Jupiter is, and there's no easily visible rocky or liquid surface and the gas gets thinner with height, where do you place the 'surface'? One commonly accepted definition of 'surface' is where the local gas pressure is the same as the pressure at sea level on Earth.

Second, some of the Hot Jupiters are so hot that they shed (or evaporate) their mass like crazy. These are Ultra Hot Jupiters, with surface temperatures up around 2,000°C and more. Their substance is just evaporating off into space! KELT-9b, with about 2.8 times the mass of Jupiter, orbits its star, HD 195689, also called KELT-9, which has a surface temperature of over 10,000°C (that's almost twice as hot as our Sun). KELT-9b loses 10,000 to 10 million tonnes of mass each second, due to the planet's incredible surface temperature of 4,300°C.

Third, here's something really weird, but true. KELT-9b could comfortably take the title of 'Hotter than Most of the Stars in the Universe'.

The 'trick' here is that about three quarters of the stars in the Universe are red dwarfs, which have a surface temperature of less than 4,000°C. KELT-9b is hotter than this.

Jupiter took some 2,000–7,000 orbits of the Sun to fully form.

Starting a bit later, and so part of the way through this process, Saturn also began to form from the same spinning disc, a little further out from the Sun.

Now, this is very unusual. In most Exoplanetary Systems not involving super-Earths, they just have one giant Jupiter-like planet, not two.

Saturn wasn't even fully formed when Jupiter began to spiral towards the Sun. Jupiter was interacting with the gas, dust and planetesimals (rocks that are 10–100 km in diameter, and bigger), and if you do the fancy physics, you can explain why Jupiter started spiralling in towards the Sun. Its trip took about 100,000 years – a very short time in planetary terms.

So what did cause this migration? It involves fancy physics, with stuff like Angular Moments and differential Lindblad Torque Transfer – you get the idea. (You'll have to do this homework by yourself.)

Location of the Sun

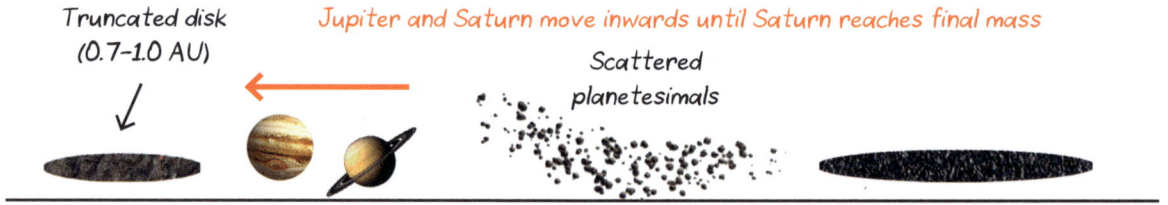

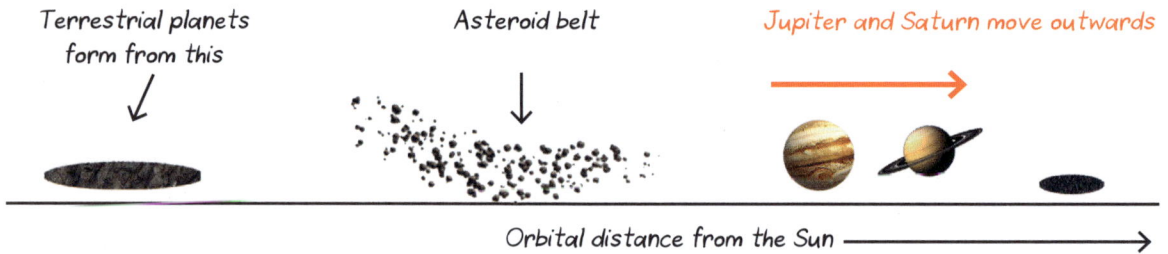

Disk of rocky planetesimals

Jupiter and Saturn migrate inwards

Disk of water- and carbon-rich planetesimals

Location of the Sun

Truncated disk (0.7–1.0 AU)

Jupiter and Saturn move inwards until Saturn reaches final mass

Scattered planetesimals

Location of the Sun

Terrestrial planets form from this

Asteroid belt

Jupiter and Saturn move outwards

Orbital distance from the Sun

But then Saturn follows the example of Jupiter, and also begins to spiral in. Because Saturn is lighter than Jupiter, it migrates much more quickly.

After just a few thousand orbits around the Sun (about 100,000 years), they end up in new temporary locations. Jupiter is around 1.5–2 AU (roughly where Mars is today), while Saturn remains a bit further out.

Jupiter and Saturn get locked into a 3:2 resonance. This means that for every three orbits of Jupiter around the Sun, Saturn does two. (As a bonus fun fact, Neptune and Pluto, one of the dwarf planets of our Solar System, are also locked into a 3:2 resonance.)

For the next 1,000-or-so orbits of the Sun, the two gas giants create a huge amount of havoc, thanks to their combined gravitational fields.

What kind of havoc? Well, there may already have been some super-Earths, orbiting close to the Sun. But the combined gravity of Jupiter and Saturn would have sent these super-Earths into what is called a 'collisional cascade'. The planetesimals (big rocks) smashed into the super-Earths, and the super-Earths might have actually crashed into each other. Either way, most of the fragments from these smash-ups fell into the Sun. The final result was that after only 20,000 years of Jupiter and Saturn sitting in their new neighbourhoods, no super-Earths were left. In fact, no rocky planets were there at all.

But something was left over from the destruction of the super-Earths – lots of gas and dust, as well as debris, remained orbiting the Sun. Over the next 30–50 million years, this material turned into the four inner small rocky planets, each with thin atmospheres.

The changes to the gas-and-dust environment meant a change to the physics of the Solar System. So both Jupiter and Saturn then migrated outwards to their current positions at 5.2 AU and 9.5 AU. This took about 500,000 years.

We think that all of this hullaballoo happened in the first few million years after the Sun began burning. I would just love to see a time-lapse movie of it.

The later stages (the formation of the four inner rocky planets) happened only because of Saturn.

Luckily for us, Saturn pulled Jupiter outwards from the Sun. If there was no Saturn, Jupiter would have remained nearer the Sun, and Earth would not have formed. We would then have been like most of those Exoplanetary Systems out there in the Milky Way galaxy.

IT EXPLAINS EVERYTHING, BUT ...

The Grand Tack hypothesis explains how the smallish inner planets formed.

The maths (which, unfortunately, is rather complicated) tells us why Mars is so light – only about 11% of the Earth's mass.

And even more complicated maths also explains how come there seem to be two types of rocks in the Asteroid Belt between Mars and Jupiter – carbon-based further out and silicon-based closer to the Sun.

However, while this Grand Tack explanation sounds so sensible and believable, just remember that old saying in physics, 'For every problem, there is a solution that is easy to understand, explains everything, and is intuitively correct – but is 100% wrong!'

Or as the scientists like to say, 'Only time and observational data confirming or rejecting testable predictions will tell' (there's no fun way to paraphrase this boring sentence). It is possible that the Grand Tack is off-track, and taking us in the wrong direction, but it's certainly an interesting hypothesis to consider.

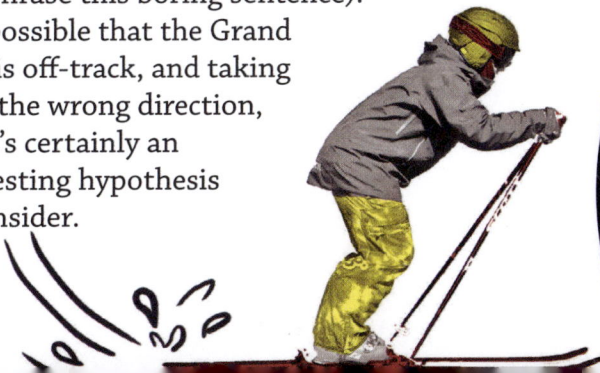

The small Mars and other problems

The standard image of the Solar System is so neat. Four little rocky planets close to the Sun, four bigger gas/ice planets further out, and a nice dividing line with the Asteroid Belt.

The fact that they're here now is taken as a kind of proof of the Solar System's stability, that they've always been there.

But …

The Sun has a lot of gravity. So, presumably, it would have attracted most of the gas and dust from the early days towards itself.

But if there was lots of gas and dust flying around in the inner Solar System, why is Mars so small? The same question also applies to Mercury.

And why are the gas planets so big, if there wasn't much stuff way out in the boondocks to make them from?

We have no definitive answers to these questions yet, but they do point towards our planets having moved around in the past, as the Grand Tack suggests.

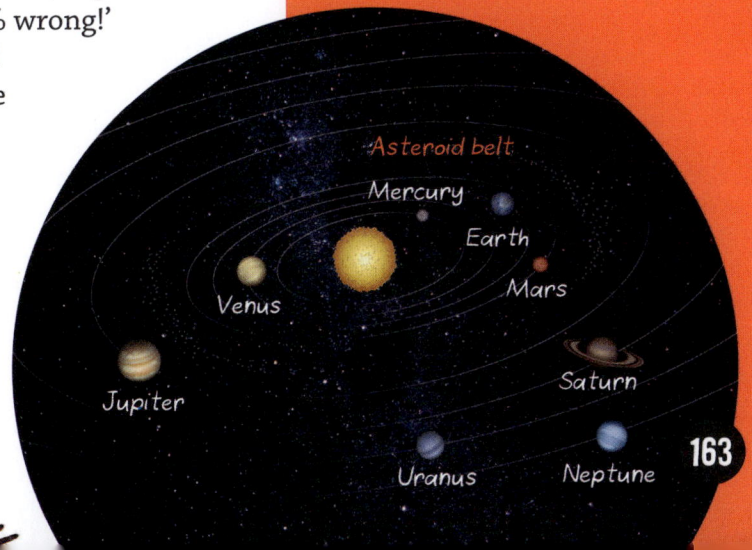

163

5G MANIA

Now, like it or not, the 5G telephone network is coming. It offers blisteringly fast download speeds – up to 20 gigabits/second (Gbps) – and reduced waiting time.

That's almost as astonishing as the abundance of negative publicity around 5G telephone networks!

The thing that is claimed to be so much more dangerous about the 5G telephone network is that it uses 'higher frequencies' than previous networks. Yes, some versions of 5G use higher frequencies – up to 10 or even 100 times higher.

But how about this? There are 'items' (I am being deliberately coy) that already broadcast radiation at frequencies some 1,000 times higher than typical 5G frequencies – and there are billions of these 'items' wherever people are. However, the electromagnetic radiation from these 'items' has never been proven to cause cancers, or any other bad health effects. What are these 'items'? See if you can guess (the answer is right at the end of this chapter).

According to the claims of the relentless rumour machine, this 'evil' high frequency 5G radiation has already caused vast numbers of various cancers in humans, as well as destroying vast swathes of forest across the world.

When these claims first came out, the 5G phone networks had been rolled out for only a few months in a handful of cities in South Korea, China and the USA. That's amazingly fast-acting, don't you think, even for 'evil' radiation?

Let's first take a look at what it means to have a 5G network.

FROM 1G TO 5G

Since the early days of mobile phones, we have moved into progressively faster networks, which have the advantage of being able to transmit our data more quickly.

The very first commercial cellular network was launched in Tokyo in 1979. That was the original 1G (first generation) network, and it was analogue. It was amazing at the time – even though all it carried was voice calls.

The 2G networks were first launched in Finland in 1991. They were very different from the 'old-fashioned' 1G analogue networks.

CODED SECRETS

Secret messages have been used by the military, and others, for thousands of years. In World War II, the Allies built a secure system that would encrypt speech (not just text, like the German Enigma system) in real time. The British mathematician Alan Turing helped design it. The system was called SIGSALY and it was enormous – it had 40 racks of hardware, chewed up about 30,000 watts of power, and weighed over 50 tonnes. Because it was so cumbersome, only about a dozen SIGSALY terminals were set up around the world – one for the Pentagon, with an extension to the White House; one in London for Winston Churchill, with an extension to the US Embassy; one on a ship, for the use of General Douglas MacArthur in his South Pacific campaign; and several others for very-high-level locations and people. SIGSALY operated from 1943 to 1946.

By the mid-1970s, corporations and governments wanted a similar ability to easily have a secure phone call, with the voice being 'encrypted' – even with an analogue network. (With the old analogue phone networks – wired or mobile – if anybody could 'tap' into a conversation, they could hear everything said by both parties.) These early stand-alone devices were relatively expensive, and complex. Each party would use this device to encrypt their voice going out, and de-encrypt the incoming voice signal coming back to them. If a third party managed to tap into the phone line, all they would hear would be indecipherable noises.

So when the first digital phone network started up, by default, it had encryption built into it from the very beginning – simply because so many corporations and governments wanted it. This early encryption was called A5/1. Unfortunately, within a few years, it was hacked. However, since then, many very powerful encryption apps/protocols (such as ZRTP) have become available. Apps using them include Silent Phone, Signal, BBM, Apple Facetime and WhatsApp (some are free). Some have encryption so powerful that even the NSA (US National Security Agency) would have difficulty busting them.

Compared with the 50-tonne 30,000-watt SIGSALY, modern phone encryption is an amazing jump in technology.

Because they were digital, the voice conversations could be easily encrypted (and they were), and you could also send and receive SMS text messages. The fastest data-transfer speeds were a paltry 40 kilobits per second (kbps) – about half a million times slower than 20 Gbps. The last 2G network shut down in Australia in mid-2018.

The 3G commercial networks were introduced in 2001. They were faster, with a minimum data-transfer rate of 144 kbps.

The 4G system was launched in Scandinavia in 2009. It enabled extra services, such as gaming, high-definition mobile TV, video conferencing, 3D television and telephone calls over the web. Obviously, all of this extra capability required higher peak data speeds – between 100 and 1,000 million bits per second (Mbps).

The 5G network is the 'latest and greatest'. It was launched in April 2019 in South Korea. It comes in several frequency bands, with potential data delivery rates up to 20 Gbps (very fast indeed, and similar to speeds found on the backbone of fibre-optic internet networks such as the National Broadband Network).

ELECTROMAGNETIC RADIATION 101

All mobile phones – including 5G – use high-frequency radio waves and microwaves. These are just two of the many types of electromagnetic (EM) waves. Other types of EM waves include light, heat, X-rays, gamma rays, etc. Heat is yet another type of EM radiation – isn't that interesting?

So let me start with an overview of the electromagnetic spectrum.

We measure the frequency of EM waves in 'cycles per second', which is usually abbreviated to 'hertz' (Hz).

The sliver of the EM spectrum that we use for vision is the visible light spectrum. It covers all the colours of the rainbow from red to violet. It sits around the middle of the very wide EM band and occupies a very tiny part of the spectrum, less than one-trillionth of one-trillionth of the entire EM band – depending on how you measure it.

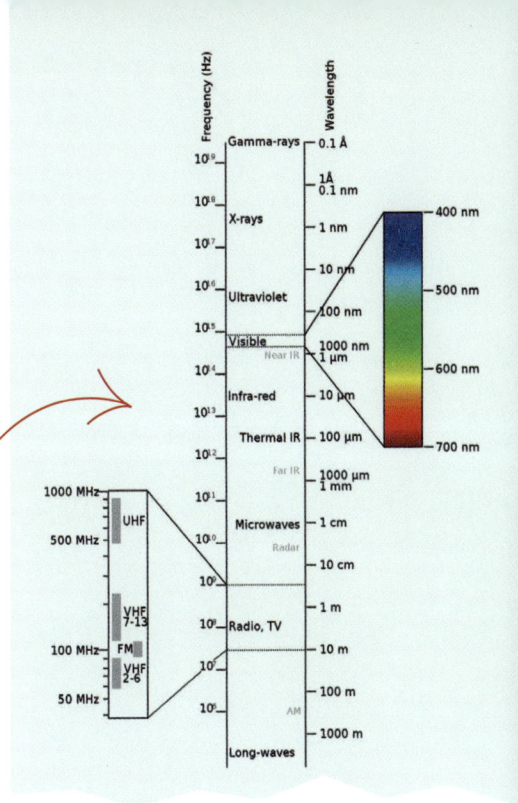

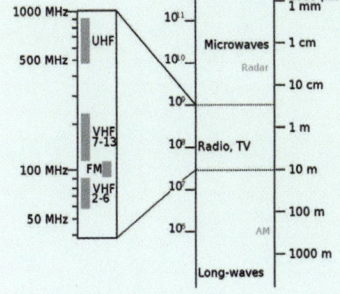

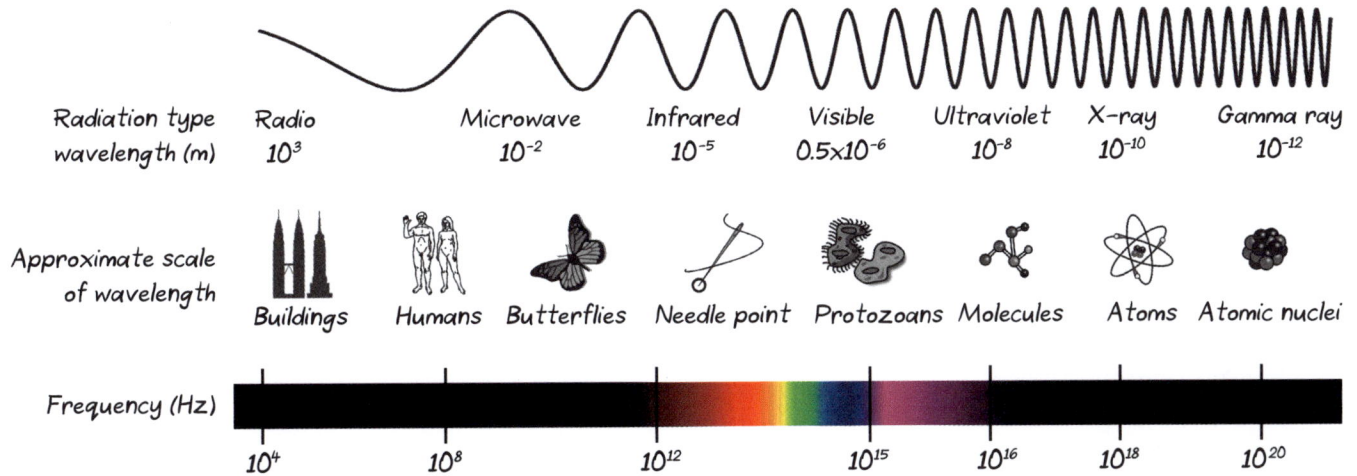

Radiation type	Radio	Microwave	Infrared	Visible	Ultraviolet	X-ray	Gamma ray
wavelength (m)	10^3	10^{-2}	10^{-5}	0.5×10^{-6}	10^{-8}	10^{-10}	10^{-12}

Approximate scale of wavelength	Buildings	Humans	Butterflies	Needle point	Protozoans	Molecules	Atoms	Atomic nuclei

Frequency (Hz) 10^4 10^8 10^{12} 10^{15} 10^{16} 10^{18} 10^{20}

Down the bottom end of the spectrum we have the extremely low frequency band, or ELF. It runs from 3 to 30 Hz (or cycles per second).

In each second, these EM waves go up and down between 3 and 30 times.

The amount of information you can send increases with the frequency. So the ELF band can send data only very slowly. But the ELF waves are hard to block – they will penetrate the ground beneath our feet, and even the oceans. So ELF frequencies can be used to communicate with submarines while they are underwater.

As we head up the spectrum to higher frequencies, we jump up to the ultra low frequency band, which runs from 300 to 3,000 Hz. Because the frequency is higher, you can send more information

in each second. The downside is that the waves don't penetrate as deeply. But they will still penetrate the ground to some degree, so they're used in underground mines.

Then as we keep moving up to the higher frequencies, we come across the million Hz (or megahertz) band, which is used for AM radio. Up a bit higher is the 100 million Hz band (or 100 megahertz), which is used for FM radio and TV.

Visible light has a much higher frequency – up around 600,000 GHz. X-rays and gamma rays have much higher frequencies again.

WHAT'S THE FREQUENCY, KENNETH?

Let's look at the frequencies mobile phones use – higher than TV, but a lot lower than visible light.

The 1G networks ran on typical radio frequencies of 150 to 850 megahertz (MHz). By the time 4G came around, the radio frequencies worldwide that carried 4G ranged from 450 to 5,900 MHz, but in Australia, they were limited to between 700 and 2,300 MHz. The 5G network can (way off in the future) potentially run at up to 300 GHz (which is 300,000 MHz).

Some people say, 'We don't need optic fibre because the new 5G network will be so fast.' But the only way that the internet can get to the 5G nodes/transmitters is via optic fibre.

The new 5G network uses three main frequency bands – low, medium and high. The data-transfer rate is much faster on these higher frequencies for two reasons. First, there are more cycles (up-and-downs, or 'waves') in each second. Second, there is lots of unused and spare bandwidth up at these higher frequencies, due to relatively few services existing up there, so there is a lot of room for very wide signals to carry lots of data.

The low, medium and high frequency bands of the new 5G have very different capabilities. The low band uses the frequency range of 600–700 MHz, similar to slow 4G. As you would expect, the data speeds are also similar – 30–250 Mbps. Many countries are not even bothering to implement this slower version.

The medium band uses radio frequencies around 2.5–3.7 gigahertz (GHz), which are similar to what microwave ovens (at much higher power levels) use to heat your food – and some 4G networks. This should allow data-transfer speeds of 100–900 Mbps. Many countries are using this as their entry-level or 'poverty-pack' 5G.

By mid-2020, only a few countries are using the high-band

3G COOKS EGGS?

Mobile networks have long been accused of causing us harm. Back in 2006, my email inbox was filled with shock-horror stories 'proving' that 3G radiation was dangerous. Here is one of them:

We need:

One egg and 2 mobiles

65 minutes to call from one phone to the other

Set up something like in the graphic

We'll initiate the call between the mobiles to last for 65 min's approximately;

Nothing will happen on the first 15 minutes...

After 25 minutes the egg starts warming up, after 45 min's;

The egg is already hot; and after 65 min's the egg is cooked

Conclusion:

If the microwave radiation emitted by the mobiles is capable to modify the proteins in the egg. Imagine what it can do with the proteins in our brains when we talk through the mobiles.

The instructions were to get two mobile phones (use one mobile phone to call the other), place the two phones face-to-face with a raw egg in between them, bind everything together with some sticky tape, and leave it all for 65 minutes. According to the email, the egg would begin to warm after about 25 minutes, and be fully cooked by 65 minutes.

The experiment was so simple that I did it with my eight-year-old daughter, Lola.

What we found was that – you guessed it – after 70 minutes, the egg between the two mobile phones had not even warmed up a fraction. I did the experiment again myself the next day, using an accurate thermometer, and the egg's temperature still hadn't budged by a single degree. Even the British TV show *Brainiac* did the experiment with 100 phones and produced the same results: no warming of the egg.

The original email was a deliberate hoax. It was started by Charlie Ivermee, an archivist at a legal firm in Southhampton in the United Kingdom, who wrote under the pen name of Suzzanna Decantworthy, from the fictional village of Wymsey. He had an electronics background, and found the whole idea of 'mobile phones frying your brain' so ridiculous that he wrote this email (with accompanying pictures) as a satirical story. To his absolute surprise, people believed it.

What's more, mobile phones don't even transmit to each other. Instead, all mobile phones transmit to a base station. So it makes no sense to have the egg between the phones – it's just a psychological thing to scare you.

Over a decade and a half later, people still send me this email. But, as Lola and I found out, anyone can easily *disprove* it in just one hour!

version of 5G (but this will change very quickly). It starts at about 6 GHz, but potentially runs up to 300 GHz. Those super high frequencies over 100GHz will require the installation of brand-new technologies and infrastructure – all the way from the nodes/transmitters to the smartphones we will hold. So at the moment, the countries using the high band are sticking to the lower end of that range – around 25–39 GHz. This allows data-transfer speeds up to 3 Gbps.

I INTRODUCED MOBILE PHONES TO AUSTRALIA!

To a very tiny (okay, insignificant) degree, I 'introduced' mobile phones to Australia.

On 3 April 1973, Martin Cooper, an engineer with Motorola, made the first ever mobile/cellular phone call. He supposedly said to his arch-rival at the opposition, Bell Labs, 'I'm ringing you just to see if my call sounds good at your end.' He used a Motorola DynaTAC mobile phone. This huge lump of a phone took ten hours to recharge its battery – and gave you just 20 short minutes of talk time. That's a poor trade.

Some 14 years later, at the Sydney Opera House at 10.42 am on 23 February 1987, Australia's first handheld mobile-phone call was made. It was a massive event. There was a big stage, with flowers and computers and loudspeakers, set up out in the open. Yes, I was the official Master of Ceremonies and host of the event.

This 'official' first mobile phone call was from the then–federal communications minister, Michael Duffy, to Mel Ward, the managing director of Telecom Australia.

The press was there, and both sides of the conversation were played over the PA system to the waiting crowd. After this first mobile-phone call came a series of further mobile-phone calls – from marathon runners carrying their heavy mobile phones while running towards the stage, from parachutists carrying their mobile phones while trying to land on the forecourt of the Opera House, and from radio personalities Jono and Dano kayaking across the harbour from the North Shore.

It was all meant to be a technological showcase. But unfortunately, some of the phone calls dropped out – something to do with the incredibly complex and cutting-edge technology, I guess.

At least, that's what I said, while I was desperately trying to cover the awkward silence. Thinking quickly, I used the empty air-time to explain the many advantages of this newfangled cellular technology. I predicted that, one day in the distant future, we could all have our own personal mobile phones. (This is one of the very few predictions I've made that have come true.)

At the time, this raised some eyebrows – after all, those phones were sardonically called 'bricks' because they were so large. They also cost more than $4,000 in 1987 dollars, and had ridiculously short talking times because of the high current drain. The 'solution' that gave you a few hours of talk time was to carry with you a heavy and bulky car battery to run the phone – which some people actually did!

SPEED VERSUS RANGE

You've probably heard the phrase 'swings and roundabouts'. It relates to fairgrounds – the owner might lose money on the swings, but will make it up on the roundabouts. Today the saying means, 'You can't have it all. You might win here, but you'll lose over there.'

And this is a pretty good way to think about the different generations of mobile phones. You can have either high data-transfer rates or range in kilometres – you can't have both.

The old 2G phones had a potential range of tens of kilometres. Their radio waves were at a low frequency, so they easily penetrated plants, buildings, water vapour, human flesh (though barely absorbed), etc. But the data-transfer rate was horrendously slow – 40 kbps.

The new 5G phones will have data-transfer rates up to 20 Gbps. But the 5G phones operate at higher frequencies, so the catch is the much shorter range over which they can send and receive the signal – usually only a few hundred metres. This means the towers used for 3G and 4G are too few in number and too far apart to fulfil the coverage requirements of 5G. In the USA, when Verizon began rolling out 5G, they found they had to place a 5G node on the corner of every city block.

On the upside, this vastly reduced range should mean that the potential health impacts of 5G will be low. Yes, at higher frequencies, radio waves penetrate less, and so they should become *more safe*, not more dangerous.

As mentioned above, the old 2G radio frequencies could penetrate your physical body – and, by the way, even then there were no adverse health effects we are aware of. 5G can barely penetrate a centimetre into your flesh, which is a very good indicator that any possible health risks will be even lower.

A QUICK RECAP

Let's recap the main points of 5G technology. (We've covered a lot of territory, so here's a chance to catch your breath.)

First, as the radio frequency gets higher, we get more cycles per second, so we can transmit more data. There's also more bandwidth (or larger bands of spectrum) up there at the higher frequencies. So it's just plain old physics that lets the 5G network transmit data to phones at typical speeds of 1 Gbps.

Second, as a frequency gets higher, the ability of the radio signal to pass through stuff reduces. Once again, plain old physics tells us that the 5G radio signal will be more easily blocked by the concrete in buildings than the signals from the old 3G or 4G phones were.

This brings us up to speed on 5G technology. So why are rumours running wild about cancer deaths being caused by mobile phone radiation?

CANCER & IONISING RADIATION

So what about the big cancer scare? What's all the talk by the conspiracy theorists about 'damaged' atoms? It turns out that 'damaged' atoms can cause cancers. But 5G doesn't damage atoms.

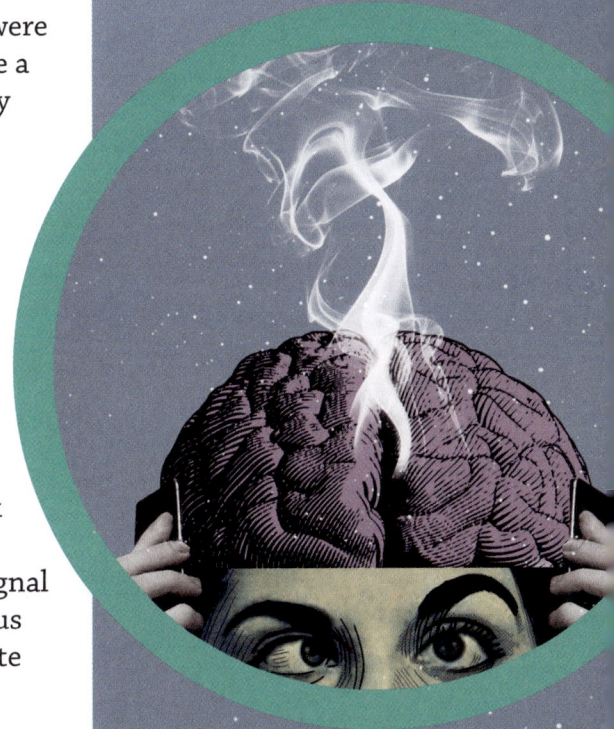

It has been long known that some electromagnetic radiation definitely can damage atoms. To be specific, the radiation knocks one (or more) of the electrons off the atoms. An atom that has lost electrons is called an 'ion', so this radiation is called 'ionising radiation'. (Just to be perfectly clear, the chemists and physicists generally call these changed atoms 'ions', rather than 'damaged atoms'.)

These ions are generally more able to carry out chemical reactions than regular atoms. If the ions react with chemicals (such as fats, proteins and carbohydrates), any damage is just purely local and usually doesn't cause a cancer. But if the ion reacts with your DNA, it can change it – which can sometimes lead to a cancer.

Ionising radiation is well known to directly cause cancer. It has enough energy to tear apart the chemical bonds inside the DNA of our cells and cause them to mutate, thus resulting in cancer. That's nasty. We saw that in the cancers caused by the bombing of Hiroshima and Nagasaki, and the Chernobyl disaster.

So that's two pathways by which ionising radiation can cause cancer – either indirectly via reactive ions when they damage the DNA, or directly by damaging chemical bonds that hold the DNA molecule together.

Quite conveniently, the border between 'ionising radiation' and 'non-ionising radiation' is just above the visible light band. And yes, this is the difference between cancer-causing, or carcinogenic, and non-carcinogenic radiation.

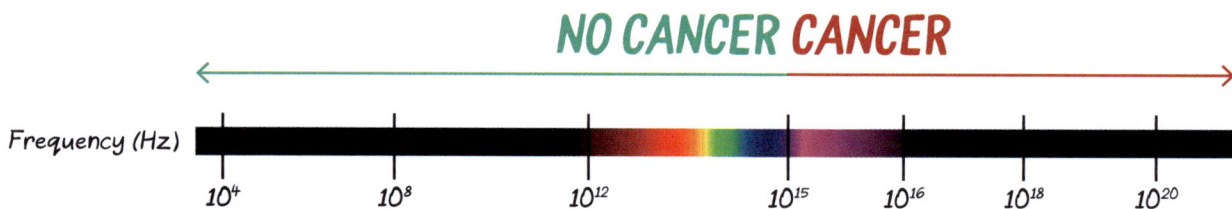

NO CANCER CANCER

Frequency (Hz)

10^4 10^8 10^{12} 10^{15} 10^{16} 10^{18} 10^{20}

You can shine violet light on skin all day and cause no damage. A photon of violet light doesn't have quite enough energy to knock electrons off atoms. It doesn't matter how many trillions of photons come together – the brighter the light, the more photons there are – they still will not knock electrons off atoms to create ions.

But if you go to a slightly higher frequency, 'ultraviolet', you have just crossed the border. All radiation with a frequency higher than violet light can cause cancer of some kind. (By the way, the very highest frequency proposed to be used by 5G, 300 GHz, is still about 10,000 times lower than the lowest frequency of ionising

5G AND COVID-19 CONSPIRACIES

During the early days of the COVID-19 pandemic, conspiracy theories circulated claiming that the 5G network was the true cause of the disease, not the virus SARS-CoV-2. As a result, Telecom masts (supposedly transmitting 5G) were set on fire in nine European countries. (Of course, the actual metal structure didn't burn. Petrol in an open container was set on fire next to the cable carrying the signal to the top of the tower. The polyethylene covering the cable ignited, stopping the transmission and reception of radio signals.) However, most of these towers were for Emergency Services, and for 3G and 4G signals. So in most cases, the arsonists actually burnt the wrong target.

In March 2020, I was emailed some exciting new conspiracy theories. They were rather ingenious. They claimed that a coronavirus vaccine, which they said had been recently released, secretly carried tiny microchips. These microchips were small enough to both be invisible to the human eye and to be injected into the flesh.

One version of this vaccine-related conspiracy theory said these were Chinese microchips that would enable some off-shore mastermind to control our thoughts via the nearest 5G transmitter.

Another version claimed that the COVID-19 disease was not caused by any coronavirus at all. Instead, all of the symptoms (ranging from being completely asymptomatic, to a mild upper respiratory tract infection, to difficulty in breathing, and finally a cytokine storm causing death) were caused entirely by these secret microchips in the vaccine. According to this theory, if the local government doesn't like you, they can easily activate the microchip when you happen to pass a 5G tower – and 20 days later, you will be dead.

A big, obvious flaw to all of these theories is that when they began to spread, in March 2020, there was (sadly) no such thing as a coronavirus vaccine. Pretty obviously, no coronavirus vaccine means no microchips in our bodies!

Other theories linking 5G and COVID-19 claim that:
• 5G weakens your immune system, making the COVID-19 symptoms very much worse (incorrect);
• the COVID-19 lockdowns around the world are just a cover to quickly install more 5G networks, especially in schools (incorrect);
• somehow, Bill Gates (from Microsoft) is the cause of 'it all' (whatever 'it all' is) (incorrect).

And now for the Big One: the 5G network is all part of an Illuminati mass-murder plot, with the 'accidental' helicopter deaths of the basketballer Kobe Bryant and his daughter being the essential satanic Blood Sacrifice needed to let the coronavirus spread around the world. (Yup, incorrect again.)

radiation. So 5G simply can't cause cancer – because it's not ionising radiation.) A photon of ultraviolet (UV) light, at a frequency of 30 million GHz, is ionising radiation – it has enough energy to blast electrons off atoms. In the real world, we all know that UV light, which makes up about 3% of the sunlight that reaches the ground, can cause skin cancers, which is why we put on sunblock.

As we keep rising up through the EM spectrum to higher and higher frequencies, we reach X-rays and finally gamma rays. The photons of these ionising radiations carry much more energy than UV. If they land on human flesh, both X-rays and gamma rays can cause cancer – and we have known this for a long time. That's why X-rays are taken only if it's really necessary for diagnosis – and why the people who operate the X-ray machinery wear lead coats and stand behind lead shields for protection.

It has been claimed that 'radiation is cumulative'. Mmmmm.

Let's look at 'ionising radiation'. Yes, a single chest X-ray has a microscopic chance of causing a cancer in your chest. And if you have a million chest X-rays, you increase that microscopic chance by a million times – and suddenly you have a significant chance of getting a cancer in your chest.

But it's completely different with 'non-ionising' radiation. A single exposure to non-ionising radiation has a zero chance of causing cancer. If you have a million exposures to that non-ionising radiation, you multiply that zero chance by a million – which is still equal to zero!

Maybe we should say that 'the effects of ionising radiation are cumulative, but this is not the case for non-ionising radiation'.

NON-IONISING RADIATION

But let's stop and change direction, away from high EM frequencies and Cancer Land, to lower frequencies and Non-Cancer Land. We arrive at some very familiar turf near the middle of the EM spectrum – the visible light band.

As we keep travelling down towards lower frequencies, we drop from violet light to red, and then into heat (which we also call near infrared, and further down, far infrared). All of these EM radiations are non-ionising. A photon of this non-ionising EM radiation does not carry enough energy to damage atoms – so it cannot cause cancer.

Visible light, AM and FM radio, TV, microwaves, mobile phones and power lines cannot cause cancer. Why? Because at these lower frequencies, the energy in each photon is just too low to damage atoms.

Scientists have run many thousands of studies over the last half-century, and we have never been able to prove that any of these non-ionising radiations cause cancer.

The World Health Organization said, writing about the EM

radiation from mobile phones, that it found 'no obvious adverse effect of exposure to low-level [EM radiation]'. The US National Cancer Institute wrote, 'No consistent evidence for an association between any source of non-ionizing EMF and cancer has been found.'

OF RATS, MICE & MEN

But what about the two oft-quoted major studies – one done with rats, the other with mice – relating to non-ionising radiation and cancer, which were released by the National Institutes of Health in the USA, in November 2018? The ones that are mentioned every time the supposed link between cancer and mobile phones comes up?

Well, one of the main findings they produced was that mobile phone radiation actually *increased* life expectancy! (In other words, the actual finding is NOT what is quoted by those worried by the radiation associated with mobile phones. I strongly suspect that the people who quote these studies have not actually read them.) However, it's important to understand that, scientifically, these were not good studies. The results were not consistent; they contained internal contradictions, and some of the findings were (in the words of the authors) statistically 'uncertain'.

Also, keep in mind that one study specifically used Sprague Dawley rats, because they are very prone to getting spontaneous cancers. In these rats, the rates of spontaneous cancers are about 22% in females, and 5% in males. By the way, note that the females are 4 to 5 times more likely to get these cancers. Spontaneous cancers are cancers that arise spontaneously, without a cancer-causing agent being involved – such as skin cancers in those never exposed to sunlight, or lung cancers in those who never smoked cigarettes. The advantage of using cancer-prone rats is that if a suspected carcinogen does not give them cancers, then you can fairly confidently say that it won't cause cancers in regular rats. Mind you, it's a whole extra jump to extrapolate that to apply to humans.

The rat study (which is described over 380 pages) exposed rats to the EM radiation put out by mobile phones running at the relatively low frequency of 900 MHz (at the lower end of 4G networks). Some 180 male and female rats, in each of several groups, were exposed to this radiation over their whole body, not just their heads. The radiation levels were very much higher than a human would get from their mobile phone. The rats were exposed for nine hours a day, seven days a week, for two continuous years.

GREEN ROOM

5G and oxygen conspiracy

One 5G conspiracy is that 5G radiation will make oxygen unavailable to our bodies! Like all good conspiracy theories, it uses a tiny (and irrelevant) element of science.

At the moment, the 5G frequencies stop at around 30 GHz. But in the future, they will be used up to 300 GHz. However (here's some genuine science), frequencies around 60 GHz will not be used.

The reason is that oxygen (20% of the atmosphere) absorbs these radio frequencies. If 5G were to use these frequencies, the radio signal would be much weaker, because some of its energy will have been lost to the oxygen. Of course, following the well-known Law of Conservation of Energy, oxygen molecules would get microscopically warmer – emphasis on 'microscopically'.

So major 5G networks will specifically not broadcast around 60 GHz. It makes no sense to broadcast expensive radio power, if it won't all get to the user.

This doesn't stop the conspiracy theorists claiming that 5G will be broadcast at 60 GHz (which it won't be), and will somehow interfere with haemoglobin's ability to carry oxygen around the human body (zero proof).

Surprisingly, the male rats that were exposed to the radiation actually lived longer than the non-exposed rats. However, they did have more cancers of the heart and brain – but weirdly, this was only for the male rats. The female rats did not have more cancers. Why, when the female rats were more prone to cancers generally, and especially when exposed to agents that cause cancer?

So with huge doses of non-ionising radiation, all the Sprague Dawley rats lived longer, but the males had more cancers (which, by the way, didn't shorten their life span).

The other study (258 pages) looked at mice. It exposed several groups of 180 male and female mice to a higher frequency used by mobile phones – 1,900 MHz.

Again, they were exposed to very high levels of radiation, for approximately nine hours a day, seven days a week, for two continuous years. And again, the male mice that were exposed to radiation lived longer than the mice that were not exposed.

In this second study, none of the male or female mice had clear evidence of increased cancers.

So let's summarise both studies. If you expose rodents to massive whole-body doses of mobile phone radiation, then:

1) both rats and mice live longer;
2) male rats get more cancers;
3) female rats don't get more cancers;
4) male and female mice do not get more cancers.

Now let's dive a little deeper into these studies.

In the first study with the cancer-prone rats, why was there a difference between the males and the females?

Almost certainly, it's because the numbers of rats with cancer were all very low – all in the single digits. This is an incredibly small sample size of positive results.

According to the neurologist Dr Steven Novella, 'This makes subtle confounding effects and also random quirky effects more likely ... The fact that the data was negative in female rats, in male and female mice, and for most tumour types is important. It limits the applicability of the results, and suggests they may be just random noise or due to some confounding factor.'

So the results are still fuzzy – we need a bigger sample size and better studies. But, on average, the radiation-exposed mice and rats did live longer.

However, these flawed studies of *rodents* somehow still get misinterpreted as showing that the radiation from mobile phones causes cancer in *humans* and reduces *human* life expectancy.

Housing near high-voltage power lines

Several of my professional colleagues (health workers, physicists, etc) have deliberately gone out of their way to buy a house near a high-voltage (hundreds of thousands of volts) electrical power transmission corridor.

First, it's always cheaper, because so many people wrongly think that the high voltage is a health risk. Second, there are fewer neighbours. Third, they have a lovely manicured lawn on the other side of the back fence – and the lawn is a few hundred metres wide and runs for kilometres. All they have to do is put a gate into the back fence ...

RUSSIA TODAY – RT

Where is all this fake news about mobile phones and 5G coming from?

Surprisingly, according to *The New York Times*, one major source of disinformation (or misleading information that is supplied intentionally) about the 5G network has been the Russian TV network that is simply called RT.

RT stands for Russia Today, and it is available worldwide. In the USA, even though it's not as popular as Fox News on TV, which gets over 2 million viewers each day, it is the most watched news outlet on YouTube – about one million views each day.

RT claims that the 5G network is linked in humans to 'brain cancer, infertility, autism, heart tumours and Alzheimer's disease'. There is *zero* scientific proof of this – especially considering that the 5G network has been running only in relatively few locations, and only since April 2019.

The RT network consistently runs 5G segments with titles such as 'A dangerous experiment on humanity', '5G apocalypse', 'Could 5G put more kids at risk for cancer?', '5G tech is "crime under international law"', and '"Totally insane": Telecom industry ignores 5G dangers'.

And where does RT get its data from? Certainly not from Hard Science.

The *Guardian* reporter Tim Dowling wrote that, on RT, 'fringe opinion takes centre stage. Reporting is routinely bolstered by testimony from experts you have never heard of, representing institutions you have never heard of.'

And here's an unexpected twist. Overseas, outside of Russia, this official Russian TV network spreads claims that there are massive health risks from the 5G network. But back inside Russia, the president, Vladimir Putin, is a very big promoter of the 5G network.

President Putin has said, 'We need to look forward. The challenge for the upcoming years is to organise universal access to high-speed internet, to start operation of the fifth-generation communication systems.'

So why is Russia doing this? According to Ryan Fox, the chief operating officer of New Knowledge, a company that tracks disinformation, 'It's economic warfare. Russia doesn't have a good 5G play, so it tries to undermine and discredit ours.' Peter Pomerantsev, the author of a book on Kremlin disinformation, *Nothing Is True and Everything Is Possible*, wrote, 'It's all about seeding lack of trust in government institutions.'

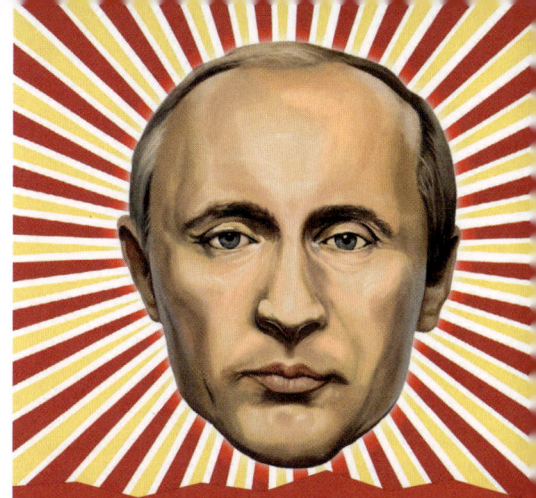

Disinformation from Russia Today 'RT'

Before Vladimir Putin became President of Russia, he served in the KGB, the Soviet Union's major intelligence agency, from 1975 to 1991. His duties in Foreign Intelligence specified that he had to spend one quarter of his time planning and carrying out disinformation.

The mid-1980s KGB disinformation campaign stating that the US Military had developed AIDS as a racial bioweapon to kill black US citizens was fabulously successful. Even as recently as 2018, a study by the University of California at Los Angeles found that half the gay black men surveyed believed this untruth.

Putin founded Russia Today in 2005. Since then, some of the deliberate disinformation it has broadcast includes:

- that the 2009 H1N1 flu was an American bioweapon;
- that the 2014 Ebola virus was an American bioweapon;
- that vaccines cannot be trusted to be safe, or to work.

Many other disinformations about bioengineered genes, radio waves, industrial chemicals, etc.

But in a very interesting twist, the medical establishment in Russia claims that the high frequencies of 5G actually make you healthier. These high-frequency therapies have been used by millions of Russian patients supposedly to treat cancer, speed the healing of wounds and boost the immune system.

'PROTECTION' AGAINST 5G

Let's get closer to home and look at a specific subset of people in our society – the folk who are anti-technology, anti-sunscreen and anti-vaccine. It turns out that they also push this anti-5G agenda – but with a twist.

They will sell you machines that will generate a supposedly 'good' electromagnetic radiation to protect you from the supposedly 'bad' electromagnetic radiation – with prices ranging from $299 to $999.

If this is not your bag, some recommend 'yoga, meditation, chanting mantras and other forms of prayer …'

And just to cover all options, they also recommend 'spirulina, wheat grass, vitamin C and similar supplements, [which] are consumable forms of sunlight, which will always improve our health and raise our vibrations'. Just what frequency they raise our vibrations to is not clear. Let's hope it's in the non-ionising range.

BILLIONS OF ITEMS THAT EMIT MORE RADIATION THAN 5G?

You might recall my claim way back, so long ago, at the beginning of this story. I reckoned that, scattered wherever people are located, there are billions of 'items' that already broadcast radiation at frequencies some 1,000 times higher than typical 5G frequencies.

These 'items' are human beings.

It's just physics and numbers – and they don't lie. We humans generate about 100 watts of waste heat. Heat is electromagnetic radiation. Our body temperature is about 37°C, which is 310.15K (K stands for the Kelvin scale of temperature measurement, which starts at absolute zero). The wavelength of this human heat is about 9,342 nanometres (nm), which is about 30,000 GHz (or 30 THz, where 1 THz, or TeraHertz = 1,000 GHz). For comparison, the wavelengths of visible light range from 400 nm (blue, 790 THz) to

POINT BREAK

5G AND WEATHER CONSPIRACY

It has been claimed that the 5G network will stop our essential weather satellites from detecting water vapour, which is an essential part of weather forecasting.

Again, there's a tiny element of science here, which is totally overrun by huge amounts of disinformation and misunderstanding.

Let me explain how weather satellites work out where the water is down here on Earth.

The water molecule H_2O emits a weak signal at 23.8 GHz. (In fact, every atom in the Universe that is above absolute zero in temperature emits some kind of electromagnetic radiation.) So weather satellites in orbit monitor the strength of the radio signal coming up from the surface at 23.8 GHz. If they see a spike in the strength of the signal at 23.8 GHz at a certain location on the planet, it indicates the presence of water vapour. The water can be breathed out by foliage on the ground, or it can be from falling rain, or any other reason – but the signal at this frequency tells us that water vapour is there. It's essential for some weather satellites to be able to monitor a clean signal at 23.8 GHz – ideally coming only from water.

But 23.8 GHz falls within 5G's potential bandwidth, which runs from 6 up to 300 GHz. Can you see the problem? What if there's a signal at 23.8 GHz coming up from a 5G transmitter? The weather satellite can't distinguish if the 23.8 GHz signal is coming from water or from a 5G antenna.

And here's another potential problem. Even if the 5G transmitters specifically do not transmit at exactly 23.8 GHz, there might be spurious 'spill-over' emissions from a nearby frequency, if an emitting device is badly made. For example, the United States Federal Communications Commission is thinking of selling some radio bandwidth to private companies at a frequency of 24.25 GHz, which is only 450 MHz away from 23.8 GHz. This has some meteorologists worried.

But the radio communications engineers are not even slightly worried. After all, 450 MHz is a big chunk of radio real estate. It's the width of 450 AM broadcast radio bands, over 22 FM broadcast bands, or more than 64 digital TV signals.

There are at least two solutions to this issue.

First, it's very easy to make filters that can block everything except 23.8 GHz. Simply install these filters on all future weather satellites.

But what about the weather satellites that have already been launched, and that have another decade or so of remaining life?

The second solution also fixes that. Again, it's very easy (and in fact, desirable) to design 5G antennas that send their power outwards parallel to the ground, with virtually no power going straight up into the sky (where the weather satellites are). This is called 'beam forming'. It costs money to send a precious 5G signal up towards the sky. And it's a useless thing to do, because there are no phones up there to pick it up. And anyway, the 5G signal gets absorbed by the atmosphere within a few hundred metres. There's no way it will get through the atmosphere to the weather satellites. But 'beam-forming' gives an extra level of safety.

There are a few other specific weather information frequencies that the future 5G network will have to avoid. They include 36–37 GHz (rain, snow), 50.2–50.4 GHz (air temperature) and 86–92 GHz (clouds, ice).

700 nm (red, 430 THz). But we radiate in the infrared band, at 9,342 nm. Infrared electromagnetic radiation (colloquially known as heat) ranges from 700–1,000,000 nm (or 430 THz–300 GHz).

Yes, each person on the planet radiates electromagnetic radiation, with a power output of 100 watts, at a frequency about 1,000 higher than 5G radiation. And this radiation has not given any of the 100-or-so billion people who have ever lived any form of cancer.

So next time you cuddle up with a friend, remember that it's electromagnetic radiation, not love, that is keeping you warm.

END NOTES

Coffee – Grinding the Perfect Cup

'Science may have found the secret to a better, and more sustainable, espresso coffee shot', Kristen Rogers, CNN, 22 January 2020, https://edition.cnn.com/2020/01/22/us/how-to-make-brew-better-espresso-study-scn/index.html

'The science behind crafting a perfect espresso', Adam Rogers, *Wired*, 22 January 2020, https://www.wired.com/story/the-science-behind-crafting-a-perfect-espresso/

'Systematically improving espresso: Insights from mathematical modeling and experiment', Michael I. Cameron et al., *Matter*, 4 March 2020, pp 1–18.

'Brewing a better espresso, the scientific way', Nick Carne, *Cosmos*, 23 January 2020, https://cosmosmagazine.com/mathematics/coffee-quality-that-adds-up

'Mathematicians say they've figured out how to brew a better espresso shot', Megan Schmidt, *Discover*, 23 January 2020, https://www.discovermagazine.com/the-sciences/mathematicians-say-theyve-figured-out-how-to-brew-a-better-espresso-shot

'Maths experts zero in on secret to perfect expresso', Nicola Davis, *The Guardian*, 23 January 2020, https://www.theguardian.com/food/2020/jan/22/maths-experts-secret-perfect-espresso

'Scientists have found a surprising formula for how to make great coffee', Kylie Walker, *SBS Food*, 23 January 2020, https://www.sbs.com.au/food/article/2020/01/23/scientists-have-found-surprising-formula-how-make-great-coffee

'International scientific study says your barista is making your coffee all wrong', Liam Mannix, *The Sydney Morning Herald*, 23 January 2020, https://www.smh.com.au/national/international-scientific-study-says-your-barista-is-making-your-coffee-all-wrong-20200122-p53trn.html

'How to brew a better espresso, according to science', Maria Temming, *Science News*, 27 January 2020, https://www.sciencenews.org/article/how-brew-better-espresso-according-science-chemistry

'Formula for an optimum cup of coffee could save billions', JAM, *New Scientist*, 1 February 2020, p 29.

Dead Fish Swim

'Passive propulsion in vortex wakes', D. N. Beal et al., *Journal of Fluid Mechanics*, 26 February 2006, pp 385–402.

'Numerical and experimental investigations of human swimming motions', Hideki Takagi et al., *Journal of Sports Sciences*, 17 August 2016, pp 1564–1580.

Self-Repairing Lungs

'Smoke signals in the DNA of normal lung cells', Gerd P. Pfeifer, *Nature*, 12 February 2020, pp 224, 225.

'Tobacco smoking and somatic mutations in human bronchial epithelium', Kenichi Yoshida et al., *Nature*, 13 February 2020, pp 266–272.

'Lung cells recover when smokers quit', James Bullen, *Choice Health*, March 2020, p 2.

Easter & Equinox

'Revealing the divine mathematics of a holiday', Ian Stewart, *Scientific American*, March 2001, http://www.whydomath.org/Reading_Room_Material/ian_stewart/2000_03.html

'How the Moon affects the date of Easter', Joe Rao, *Scientific American*, 6 April 2012, https://www.scientificamerican.com/article/how-the-moon-affects-the-date-of-easter/

'Are day and night equal at equinoxes?', Bruce McClure, *EarthSky*, 21 September 2019, https://earthsky.org/astronomy-essentials/why-arent-day-and-night-equal-on-the-day-of-the-equinox?mc_cid=298bbe9af2&mc_eid=f41436aedb

Spiders Can Fly

'Ballooning spiders: The case for electrostatic flight', Peter W. Gorham, arXiv:1309.4731v2 [physics.bio-ph], 19 November 2013.

'An observational study of ballooning in large spiders: Nanoscale multi-fibers enable spiders' soaring flight', Moonsung Cho et al., *PLOS Biology*, 14 June 2018, https://journals.plos.org/plosbiology/article?id=10.1371/journal.pbio.2004405

'Spiders can use electricity in the air to balloon for kilometres', Alison George, *New Scientist*, 5 July 2018, https://www.newscientist.com/article/2173544-spiders-can-use-electricity-in-the-air-to-balloon-for-kilometres/#ixzz6Fgt4E5T4

'Spiders can fly hundreds of miles using electricity', Ed Yong, *The Atlantic*, 5 July 2018, https://www.theatlantic.com/science/archive/2018/07/the-electric-flight-of-spiders/564437/

'Electric fields elicit ballooning in spiders', Erica L. Morley and Daniel Robert, *Current Biology*, 23 July 2018, pp 2324–2330.

'Soaring spiders', *National Geographic*, May 2019, pp 28, 29.

Past Plagues & Coronavirus

'Emerging infections: A perpetual challenge', David M. Morens et al., *Lancet Infectious Diseases*, November 2008, pp 710–719.

Defeating the Ministers of Death: The Compelling History of Vaccination, Professor David Isaacs, HarperCollins Publishers Australia, Sydney, 2019.

'Black plague, Spanish flu, smallpox: All hold lessons for coronavirus', Ibrahim Al-Marashi, *The Bulletin of the Atomic Scientists*, 13 March 2020, https://thebulletin.org/2020/03/black-plague-spanish-flu-smallpox-all-hold-lessons-for-coronavirus/

Red Sky at Night

'Why the sky is blue. For reals', Dr Karl Kruszelnicki, ABC Radio National, 17 October 2017, https://www.abc.net.au/radionational/programs/greatmomentsinscience/dr-karl-why-the-sky-is-blue/8855386

Color and Light in Nature, David K. Lynch and William Livingston, Cambridge University Press, 1995, pp 21–32.

'Is there scientific validity to the saying, "Red sky at night, sailors' delight; red sky in the morning, sailors take warning"?', Joe Sienkiewicz, *Scientific American*, 23 June 2003, https://www.scientificamerican.com/article/is-there-scientific-valid/

How To: Absurd Scientific Advice for Common Real-World Problems, Randall

Munroe, Riverhead Books, New York, 2019, pp 132–135.

Black Holes – Close & Missing

'What exactly is a black hole horizon (and what happens there)?', Charles Q. Choi, Space.com, 9 April 2019, https://www.space.com/black-holes-event-horizon-explained.html

'A naked-eye triple system with a nonaccreting black hole in the inner binary', Thomas Rivinius et al., *Astronomy & Astrophysics*, 6 May 2020, https://doi.org/10.1051/0004-6361/202038020

'Physicists discover nearest black hole to Earth – so far', Tami Freeman, *Physics World*, 6 May 2020, https://physicsworld.com/a/astronomers-discover-nearest-black-hole-to-earth-so-far/

'Astronomers find closest black hole to Earth, hints of more', The Associated Press, *The New York Times*, 6 May 2020, https://www.nytimes.com/aponline/2020/05/06/science/ap-us-sci-black-hole-nearby.html

'Astronomers may have found the closest black hole to Earth', Jonathan O'Callaghan, *Scientific American*, 6 May 2020, https://www.scientificamerican.com/article/astronomers-may-have-found-the-closest-black-hole-to-earth/

'Closest black hole to Earth has two partners in surprising celestial marriage', Reuters, *The New York Times*, 6 May 2020, https://www.nytimes.com/reuters/2020/05/06/world/europe/06reuters-space-exploration-blackhole.html

'Astronomers discover black hole closer to Earth than any before, suggesting it may not be alone', ABC/AP, ABC News, 7 May 2020, https://www.abc.net.au/news/2020-05-07/astronomers-discover-closest-black-hole-to-earth-ever/12222154

'You can "see" the closest known black hole to Earth with the naked eye', Leah Crane, *New Scientist*, 6 May 2020, https://www.newscientist.com/article/2242741-you-can-see-the-closest-known-black-hole-to-earth-with-the-naked-eye/

Coronavirus & Copper

'Copper could help prevent the spread of flu infections', *Science Daily*, 14 February 2006, https://www.sciencedaily.com/releases/2006/02/060214080834.htm

'Bacterial killing by dry metallic copper surfaces', Christophe Espirito Santo et al., *Applied and Environmental Microbiology*, February 2011, pp 794–802.

'Metallic copper as an antimicrobial surface', Gregor Grass et al., *Applied and Environmental Microbiology*, March 2011, pp 1541–1547.

'Copper is great at killing superbugs – so why don't hospitals use it?', Bill Keevil, *The Conversation*, 25 February 2017, https://theconversation.com/copper-is-great-at-killing-superbugs-so-why-dont-hospitals-use-it-73103

'Copper kills coronavirus. Why aren't our surfaces covered in it?', Mark Wilson, *FastCompany*, 16 March 2020, https://www.fastcompany.com/90476550/copper-kills-coronavirus-why-arent-our-surfaces-covered-in-it

'Aerosol and surface stability of SARS-CoV-2 as compared with SARS-CoV-1', Neeltje van Doremalen et al., *New England Journal of Medicine*, 17 March 2020, pp 1–3.

'COVID-19: How long does the coronavirus last on surfaces?', Richard Gray, *BBC Future*, 17 March 2020, https://www.bbc.com/future/article/20200317-covid-19-how-long-does-the-coronavirus-last-on-surfaces

'Here's how long the coronavirus will last on surfaces, and how to disinfect those surfaces', Yasemin Saplakoglu, *LiveScience*, 18 March 2020, https://www.livescience.com/how-long-coronavirus-last-surfaces.html

'How long will coronavirus live on surfaces or in the air around you?', Apoorva Mandavilli, *The New York Times*, 20 March 2020, https://www.nytimes.com/2020/03/17/health/coronavirus-surfaces-aerosols.html

'Copper's virus-killing powers were known even to the ancients', Jim Morrison, *The Smithsonian*, 14 April 2020, https://www.smithsonianmag.com/science-nature/copper-virus-kill-180974655/

The Amazing Disappearing Anus

The Art of Being a Parasite, Claude Combes, The University of Chicago Press, 2005, pp 21–33.

'Scorpion sheds "tail" to escape: Consequences and implications of autotomy in scorpions (Buthidae; Ananteris)', Camilo I. Mattoni et al., *PLOS One*, 28 January 2015, DOI:10.1371/journal.pone.0116639.

'How the scorpion lost its tail (and its anus)', Ed Yong, *National Geographic*, 29 January 2015, https://www.nationalgeographic.com/science/phenomena/2015/01/29/how-the-scorpion-lost-its-tail-and-its-anus/

'Getting to the bottom of anal evolution', Andres Hejnol and Jose M. Martin-Duran, *Zoologischer Anzeiger*, 27 February 2015, http://dx.doi.org/10.1016/j.jcz.2015.02.006

'The origin of the anus', Matt Walker, 11 March 2015, BBC Earth, https://www.bbc.com/earth/story/20150313-the-origin-of-the-anus

'Comb jelly "anus" guts ideas on origin of through-gut', Amy Maxmen, *Science*, 25 March 2016, pp 1378–1380.

'No surprise that comb jellies poop', Sidney L. Tamm, *Science*, 3 June 2016, p 1182.

'Defecation by the ctenophore *Mnemiopsis leidyi* with an ultradian rhythm through a single transient anal pore', Sidney L. Tamm, *Invertebrate Biology*, 22 February 2019, pp 3–16.

'This animal's anus only exists when it needs to poop, and we have so many questions', Mike McRae, *Science Alert*, 6 March 2019, https://www.sciencealert.com/this-animal-s-butt-appears-when-needed-and-it-could-help-us-understand-how-ours-evolved

'Meet the sea creature with an anus that disappears when not in use', CBC Radio, 7 March 2019, https://www.cbc.ca/radio/asithappens/as-it-happens-thursday-edition-1.5046925/meet-the-sea-creature-with-an-anus-that-disappears-when-not-in-use-1.5046932

'Discovery of animal with vanishing anus ends 160-year game of hide-and-seek', Yasmin Tayag, *Inverse*, 3 August 2019, https://www.inverse.com/article/53846-transient-anus-warty-comb-jelly

'The animal with an anus that disappears', Micheal Le Page, *New Scientist*, 9 March 2019, p 8.

The Invertebrate Tree of Life, Gonzalo Giribet and Gregory D. Edgecombe, Princeton University Press, 2020.

Murder Hornets – Lethal but Tasty?

'"Murder hornets" in the US: The rush to stop the Asian giant hornet', Mike Baker, *The New York Times*, 2 May 2020, https://www.nytimes.com/2020/05/02/us/asian-giant-hornet-washington.html

'Murder hornets vs honeybees: A swarm of bees can cook invaders alive', Mike Baker, 3 May 2020, https://www.nytimes.com/2020/05/03/us/murder-hornets-asian-giant-hornet-bees.html

'"Murder hornets", with sting that can kill, land in the USA', The Associated Press, *The New York Times*, 4 May 2020, https://www.nytimes.com/aponline/2020/05/04/us/ap-us-murder-hornets.html

'In Japan, the "murder hornet" is both a lethal threat and a tasty treat', Ben Dooley, *The New York Times*, 5 May 2020, https://www.nytimes.com/2020/05/05/world/asia/murder-hornet-japan.html

'Just how dangerous is the "murder hornet"?', Paige Embry, *Scientific American*, 6 May 2020, https://www.scientificamerican.com/article/just-how-dangerous-is-the-murder-hornet/

'Bug experts dismiss worry about US "murder hornets" as hype', The Associated Press, *The New York Times*, 7 May 2020.

'Do "Murder Hornets" Really Exist?', Matt Alt, *The New Yorker*, 13 May, 2020, https://www.newyorker.com/culture/culture-desk/do-murder-hornets-really-exist

Women's Work – Never Done, Never Paid

'Abolish billionaires', Farhad Manjoo, *The New York Times*, 6 February 2019, https://www.nytimes.com/2019/02/06/opinion/abolish-billionaires-tax.html

'Time to care: Unpaid and underpaid care work and the global inequality crisis', Oxfam Briefing Paper, January 2020, https://oxfamilibrary.openrepository.com/bitstream/handle/10546/620928/bp-time-to-care-inequality-200120-en.pdf

'World's billionaires have more wealth than 4.6 billion people', Oxfam International Press Release, 20 January 2020, https://www.oxfam.org/en/press-releases/worlds-billionaires-have-more-wealth-46-billion-people

'Women's unpaid labour is worth $10,900,000,000,000', Gus Wezerek and Kristen R. Ghodsee, *The New York Times*, 5 March 2020, https://www.nytimes.com/interactive/2020/03/04/opinion/women-unpaid-labor.html

Black Holes Have No Size

'First-ever picture of a black hole unveiled', Nadia Drake, *National Geographic*, 10 April 2019, https://www.nationalgeographic.com/science/2019/04/first-picture-black-hole-revealed-m87-event-horizon-telescope-astrophysics/

'Here's the first picture of a black hole', Lisa Grossman and Emily Conover, Science News for Students.org, 10 April 2019, https://www.sciencenewsforstudents.org/article/black-hole-first-photo-event-horizon-telescope

Pterosaurs, Not Just Pterodactyls

'Tiniest perching pterosaur discovered', Bob Holmes, *New Scientist*, 11 February 2008, https://www.newscientist.com/article/dn13298-tiny-perching-pterosaur-discovered/

'Discovery of a rare arboreal forest-dwelling flying reptile (Pterosauria, Pterodactyloidea) from China', Xiaolin Wang et al., Proceedings of the National Academy of Sciences, 12 February 2008, pp 1983–1987.

'Models show pterosaurs flew long, slow', Abbie Thomas, ABC Science, 24 November 2010, http://www.abc.net.au/science/articles/2010/11/24/3074271.htm

'Flight in slow motion: aerodynamics of the pterosaur wing', Colin Palmer, Proceedings of the Royal Society B, 24 November 2010, pp 1881–1885.

'Flamingo-like pterosaur used gravel for digestion', *New Scientist*, 14 June 2013, https://www.newscientist.com/article/mg21829215-000-flamingo-like-pterosaur-used-gravel-for-digestion/

'Pterosaurs cast a giant shadow over today's biggest winged creatures', Brian Palmer, *The Guardian*, 18 April 2014, https://www.theguardian.com/science/2014/apr/18/pterosaurs-evolution-giant-flying-animals

'3-D fossil eggs reveal how pterosaurs lived', Will Dunham, ABC Science, 6 June 2014, https://www.abc.net.au/science/articles/2014/06/06/4020263.htm

'Sexually dimorphic tridimensionally preserved pterosaurs and their eggs from China', Xiaolin Wang et al., *Current Biology*, 16 June 2014, pp 1–8.

'Giants of the sky', Daniel T. Ksepka and Michael Habib, *Scientific American*, April 2016, pp 64–71.

'Late Maastrichtian pterosaurs from North Africa and mass extinction of Pterosauria at the Cretaceous–Paleogene boundary', Nicholas R. Longrich et al., *PLOS Biology*, 13 March 2018, https://doi.org/10.1371/journal.pbio.2001663

'Monsters of the Mesozoic skies', Michael B. Habib, *Scientific American*, October 2019, pp 26–33.

Marathon Runners' Gut Bacteria

'Revised Estimates for the Number of Human and Bacterial Cells in the Body', Ron Sender et al., *PLOS Biology*, 19 August 2016, https://www.ncbi.nlm.nih.gov/pmc/articles/PMC4991899/pdf/pbio.1002533.pdf

'Another kind of superbug: Seeking an edge in the athlete's microbiome', Andrew Joseph and Drew Q. Joseph, Stat, 15 May 2017, https://www.statnews.com/2017/05/15/athletes-microbiome/

'Meta-omics analysis of elite athletes identifies a performance-enhancing microbe that functions via lactate metabolism', Jonathan Scheiman et al., *Nature Medicine*, July 2019, pp 1104–1109.

'Why are elite athletes different than the rest of us? Take a look at their microbes', Elizabeth Cooney, Stat, 24 June 2019, https://www.statnews.com/2019/06/24/elite-athletes-different-microbes/

'Elite athletes' gut bacteria give rodent runners a boost', Emily Willingham, *Scientific American*, 24 June 2019, https://www.scientificamerican.com/article/elite-athletes-rsquo-gut-bacteria-give-rodent-runners-a-boost/

'Elite runners' microbes make mice mightier', Karen Hopkins, *Scientific American*, 24 June 2019, https://www.scientificamerican.com/podcast/episode/elite-runners-microbes-make-mice-mightier/

'Can microbes increase exercise performance in athletes?', Sarah M. Turpin-Nolan et al., *Nature Reviews Endocrinology*, 1 November 2019, pp 629–630.

Combustion Engines – The Burners Go Bust?

'GM China doubles the number of electric vehicles in the works: 20 electrified vehicles will debut through 2023', Reuters, 5 June 2018, https://www.autoblog.com/2018/06/05/gm-china-ev-electric-cars/?ncid=edlinkusauto00000015

'China start-up readies capacity for 150,000 electric cars per year', Reuters, 13 June 2018, https://www.reuters.com/article/us-byton-evs-duerr/china-start-up-readies-capacity-for-150000-electric-cars-per-year-idUSKBN1J91RU

'Shenzhen's silent revolution: world's first fully electric bus fleet quietens Chinese megacity', Matthew Keegan, *The Guardian*, 12 December 2018, https://www.theguardian.com/cities/2018/dec/12/silence-shenzhen-world-first-electric-bus-fleet

'German industry to invest $68B in EVs, automation in next 3 years', *Automotive News*, 3 March 2019, https://www.autonews.com/automakers-suppliers/german-industry-invest-68b-evs-automation-next-3-years

'China moves to stop a crash in booming electric-car industry', Bloomberg News, 4 June 2019, https://www.bloomberg.com/news/articles/2019-06-04/china-drafting-measures-to-curb-18-billion-electric-car-bubble

'China drives into the future', Donna Lu, *New Scientist*, 13 July 2019, pp 18, 19.

'Full focus on electric drive', Gregor Hebermehl, *Auto Motor und Sport*, 17 September 2019, https://www.auto-motor-und-sport.de/tech-zukunft/daimler-stoppt-verbrennungsmotoren-entwicklung-2019/, 17 September 2019

'Daimler abandons internal combustion engine development to focus on EVs', Simon Alvarez, Teslarati, 19 September 2019, https://www.teslarati.com/daimler-abandons-internal-combustion-engine-over-evs/

'Subaru sets electric vehicles target', Kevin Buckland, *The Australian Financial Review*, 21 January 2020, p 20.

'Big rigs begin to trade diesel for electric motors', Susan Carpenter, *The New York Times*, 19 March 2020, https://www.nytimes.com/2020/03/19/business/electric-semi-trucks-big-rigs.html

'The body electric' Jon Gertner, *Wired*, April 2020, pp 68–73.

'One in two car sales in Norway electric, while petrol and diesel decline 60%', Bridie Schmidt, Bloomberg, 8 May 2020, https://www.bloomberg.com/news/articles/2019-06-04/china-drafting-measures-to-curb-18-billion-electric-car-bubble

Time is Running Out for Sand

'Sand, rarer than one thinks', UNEP Global Environmental Alert Service, March 2014, https://wedocs.unep.org/bitstream/handle/20.500.11822/8665/GEAS_Mar2014_Sand_Mining.pdf?sequence=3

'The world's disappearing sand', Vince Beiser, *The New York Times*, 23 June 2016, https://www.nytimes.com/2016/06/23/opinion/the-worlds-disappearing-sand.html

'Even desert city Dubai imports its sand. This is why', Renuka Rayasam, BBC, 6 May 2016, https://www.bbc.com/worklife/article/20160502-even-desert-city-dubai-imports-its-sand-this-is-why

'How to steal a river', Rollo Romig, *The New York Times*, 1 March 2017, https://www.nytimes.com/2017/03/01/magazine/sand-mining-india-how-to-steal-a-river.html

'The world is running out of sand', David Owen, *The New Yorker*, 22 May 2017, https://www.newyorker.com/magazine/2017/05/29/the-world-is-running-out-of-sand

'How the demand for sand is killing rivers', Harriet Constable, BBC News, 3 September 2017, https://www.bbc.com/news/magazine-41123284

'Time is running out for sand', Mette Bendixen et al., Nature, 4 July 2019, pp 29–31.

'A looming tragedy of the sand commons', Aurora Torres et al., *Science*, 8 September 2019, pp 970–971.

Vegan Diet

'It ain't easy eating greens: Evidence of bias toward vegetarians and vegans from both source and target', Cara C. MacInnis and Gordon Hodson, *Group Processes and Intergroup Relations*, 1 November 2017, pp 721–744, https://journals.sagepub.com/doi/full/10.1177/1368430215618253

'Why you should eat a plant-based diet, but that doesn't mean being a vegetarian', Katherine Livingstone, *The Conversation*, 13 July 2017, https://theconversation.com/why-you-should-eat-a-plant-based-diet-but-that-doesnt-mean-being-a-vegetarian-78470

'Why do vegans have such bad reputations?', Tani Khara, *The Conversation*, 6 November 2018, https://theconversation.com/why-do-vegans-have-such-bad-reputations-103683

'Why people become vegans: The history, sex and science of a meatless existence', Joshua T. Beck, *The Conversation*, 19 November 2018, https://theconversation.com/why-people-become-vegans-the-history-sex-and-science-of-a-meatless-existence-106410

'Should veganism receive the same legal protection as a religion?', Jonathan Seglow, *The Conversation*, 1 April 2019, https://theconversation.com/should-veganism-receive-the-same-legal-protection-as-a-religion-114243

'Here's why well-intentioned vegan protesters are getting it wrong', Tani Khara, 10 April 2019, https://theconversation.com/heres-why-well-intentioned-vegan-protesters-are-getting-it-wrong-115224

'Have you gone vegan? Keep an eye on these 4 nutrients', Clare Collins, *The Conversation*, 30 April 2019, https://theconversation.com/have-you-gone-vegan-keep-an-eye-on-these-4-nutrients-107708

'Pregnant women and babies can be vegans but careful nutrition planning is essential', Clare Collins, *The Conversation*, 1 May 2019, https://theconversation.com/pregnant-women-and-babies-can-be-vegans-but-careful-nutrition-planning-is-essential-107709

Fisher Space Pen – Out of This World!

'The Pen', *Seinfeld*, 2 October 1991, https://www.youtube.com/watch?v=V3Vm_ksWreM

'Interview with Paul Fisher, inventor of the SPACE PEN for NASA', *Audiology Online*, 12 April 2004, https://www.audiologyonline.com/interviews/interview-with-paul-fisher-inventor-1635

'Fact or fiction? NASA spent millions to develop a pen that would write in space, whereas the Soviet Cosmonauts used a pencil', Clara Curtin, *Scientific American*, 20 December 2006, https://www.scientificamerican.com/article/fact-or-fiction-nasa-spen/

'The Fisher Space Pen boldly writes where no man has written before', Jimmy Stamp, *Smithsonian Magazine*, 11 January 2013, https://www.smithsonianmag.com/arts-culture/the-fisher-space-pen-boldly-writes-where-no-man-has-written-before-1020748/

'The saga of writing in space', Caleb Wong, National Air and Space Museum, 9 June 2017, https://airandspace.si.edu/stories/editorial/saga-writing-space

Fish Exercising

'Metabolism, energetic demand, and endothermy', Kenneth J. Goldman, March 2004, DOI: 10.1201/9780203491317.ch7, pp 203–224

'The French press: a repeatable and high-throughput approach to exercising zebrafish (Danio rerio)', Takuji Usui et al., Peer J, 17 January 2018, 6:e4292; DOI 10.7717/peerj.4292

'Improbable research: Wigs in roads, fish in coffee press [continued]', *Journal of Improbable Results*, January–February 2002, p 5.

Surfing Space-Time – Einstein's Right Again!

'Bizarre cosmic dance offers fresh test for General Relativity', Lee Billings, *Scientific American*, 30 January 2020, https://www.scientificamerican.com/article/bizarre-cosmic-dance-offers-fresh-test-for-general-relativity/?utm_source=Nature+Briefing&utm_campaign=f418c61%E2%80%A6

'Two stars with an odd wobble are stretching space and time around them', Leah Crane, *New Scientist*, 30 January 2020, https://www.newscientist.com/article/2231746-two-stars-with-an-odd-wobble-are-stretching-space-and-time-around-them/#ixzz6DFgvtEnQ

'Lense-Thirring frame dragging induced by a fast-rotating white dwarf in a binary pulsar system', V. Venkatraman Krishnan et al., *Science*, 31 January 2020, pp 577–580.

'Space-time is swirling around a dead star, proving Einstein right again', Charles Q. Choi, 31 January 2020, https://www.space.com/einstein-general-relativity-frame-dragging.html

'An Isolated White Dwarf with 317-Second Rotation and Magnetic Emission', Joshua S. Reding et al., 23 March 2020, https://arxiv.org/pdf/2003.10450.pdf

Carbs & Ultra-Processed Foods

The End of Overeating: Taking Control of the Insatiable American Appetite, David Kessler, Penguin Books, Camberwell, Vic, 2009.

'Craving an ice-cream fix', Tara Parker-Pope, *The New York Times*, 20 September 2012, https://well.blogs.nytimes.com/2012/09/20/craving-an-ice-cream-fix/

'An inconvenient truth about our food', Mark Bittman, *The New York Times*, 13 May 2014, https://www.nytimes.com/2014/05/14/opinion/bittman-an-inconvenient-truth-about-our-food.html

'Ultra-processed foods and recommended intake levels of nutrients linked to non-communicable diseases in Australia: evidence from a nationally representative cross-sectional study', Priscilla P. Machado et al., BMJ Open 2019; 9:e029544. doi:10.1136/bmjopen-2019-029544

'Ultra-processed diets cause excess calorie intake and weight gain: an inpatient randomized controlled trial of ad libitum food intake', Kevin D. Hall et al., *Cell Metabolism*, 2 July 2019, pp 67–77.

'Obesity on the brain', Ellen Ruppel Shell, *Scientific American*, October 2019, pp 38–45.

Spiders Can Design, Build – And Count!

'Spatial learning affects thread tension control in orb-web spiders', Kensuke Nakata, *Biology Letters*, 23 August 2013, http://dx.doi.org/10.1098/rsbl.2013.0052

'The execution of planned detours by spider-eating predators', Fiona R. Cross and Robert R. Jackson, *Journal of the Experimental Analysis of Behaviour*, 18 January 2016, https://doi.org/10.1002/jeab.189

'Representation of different exact numbers of prey by a spider-eating predator', Fiona R. Cross and Robert R. Jackson, *Interface Focus*, 21 April 2017, https://doi.org/10.1098/rsfs.2016.0035.

'Portia's capacity to decide whether a detour is necessary', Fiona R. Cross and Robert R. Jackson, *Journal of Experimental Biology*, 7 August 2019, doi:10.1242/jeb.203463.

'Spider smarts', David Robinson, *New Scientist*, 8 February 2020, pp 42–45.

Winded – A Breathtaking Experience

'What happens when you get winded', BBC Sport, 28 September 2005, http://news.bbc.co.uk/go/pr/fr/-/sport2/hi/health_and_fitness/4275414.stm

'Here's what happens when you get the wind knocked out of you', Julia Calderone, 8 October 2015, https://

www.businessinsider.com.au/diaphragm-spasm-wind-lungs-breath-knocked-out-of-you-2015-10

Migrating Planets – A Different Tack

'Our "new improved" Solar System', Kelly Beatty, *Sky & Telescope*, 16 October 2010, https://skyandtelescope.org/astronomy-news/our-new-improved-solar-system/

'A low mass for Mars from Jupiter's early gas-driven migration', Kevin J. Walsh et al., *Nature*, 14 July 2011, pp 206–209.

'Two phase, inward-then-outward migration of Jupiter and Saturn in the gaseous Solar Nebula', Arnaud Pierens and Sean N. Raymond, arXiv:1107.5656 [astro-ph.EP], 28 July 2011, https://arxiv.org/abs/1107.5656v1

'Building terrestrial planets', A. Morbidelli et al., *Annual Review of Earth and Planetary Science*, 2012, pp 251–275.

'Jupiter's decisive role in the inner Solar System's early evolution', Konstantin Batygin and Greg Laughlin, Proceedings of the National Academy of Sciences, 23 March 2015, https://doi.org/10.1073/pnas.1423252112

'Timing of the formation and migration of giant planets as constrained by CB chrondrites', Brandon C. Johnson et al., *Science Advances*, 16 December 2016, https://advances.sciencemag.org/content/2/12/e1601658

'The partitioning of the inner and outer Solar System by a structured protoplanetarydisk', R. Brasser and S. J. Mojzsis, *Nature Astronomy*, 13 January 2020, https://www.nature.com/articles/s41550-019-0978-6

'A cosmic gatekeeper divides our Solar System in two', Yasemin Saplakoglu, *Live Science*, 14 January 2020, https://www.livescience.com/great-divide-separates-solar-system.html

'Jupiter may have destroyed early planets and paved the way for Earth', Alex Knapp, *Forbes*, 24 March 2015, https://www.forbes.com/sites/alexknapp/2015/03/24/jupiter-may-have-destroyed-early-planets-and-paved-the-way-for-earth/#611ffd7b2d5f

5G Mania

'Spontaneous tumours in Sprague-Dawley and Long-Evans rats and in their F1 hydrids: Carcinogenic effect of total-body x-irradiation', Ludwik Gross and Yolande Dreyfuss, Proceedings of the National Academy of Sciences USA, November 1979, pp 5910–5913.

'40 years on, mobile phones still pushing consumers' buttons', Asher Moses, *The Sydney Morning Herald*, 5 April 2013, https://www.smh.com.au/technology/40-years-on-mobile-phones-still-pushing-consumers-buttons-20130404-2h9ua.html

'NTP technical report on the toxicology and carcinogenesis studies in Hsd:Sprague Dawley SD Rats exposed to whole-body radio frequency radiation at a frequency (900 MHz) and modulations (GSM and CDMA) used by cell phones', National Toxicology Program, November 2018, NTP TR 595, pp 1–380.

'NTP Technical Report on the Toxicology and Carcinogenesis Studies in B6C3F1/N exposed to whole-body radio frequency radiation at a frequency (1,900 MHz) and modulations (GSM and CDMA) used by cell phones', National Toxicology Program, November 2018, NTP TR 596, pp 1–280.

'That rat cellphone study – I'm still not impressed', Steven Novella, NEUROLOGICAblog, 2 November 2018, https://theness.com/neurologicablog/index.php/that-rat-cellphone-study-im-still-not-impressed/

'5G data networks threaten forecasts', Alexandra Witze, *Nature*, 2 May 2019, pp 17, 18.

'Your 5G phone won't hurt you. But Russia wants you to think otherwise', William J. Broad, *The New York Times*, 12 May 2019, https://www.nytimes.com/2019/05/12/science/5g-phone-safety-health-russia.html

'The Russian government-funded TV network's hyperbolic campaign against US 5G', Matt Field, *Bulletin of the Atomic Scientists*, 15 May 2019, https://thebulletin.org/2019/05/the-russian-government-funded-tv-networks-hyperbolic-campaign-against-us-5g/

'5G: Upgrade or uncertainty?', Brian Dunning, Skeptoid Podcast #677, 28 May 2019, https://skeptoid.com/episodes/4677

'Ignore the scaremongers. 5G won't interfere with weather satellites. Here's why', Ed Oswald, *Digital Trends*, 29 May 2019, https://www.digitaltrends.com/mobile/5g-weather-satellite-interference/

'Debate rages over 5G impact on US weather forecasting', Peter Gwynne, *Physics World*, 31 May 2019, https://physicsworld.com/a/debate-rages-over-5g-impact-on-us-weather-forecasting/

'Space for peace: Not 5G or 6G', Duncan, *Nexus*, June–July 2019, p 2.

'Preparing for 5G: With 5G being rolled around the world, now is the time to be equipped', Advertisements, *Nexus*, June–July 2019, pp 6, 7.

'5G: The big picture', Jeremy Naydler, *Nexus*, June–July 2019, pp 17–25.

'Does 5G pose health risks?', Reality Check Team, BBC News, 15 July 2019, https://www.bbc.com/news/world-europe-48616174

'The 5G health hazard that isn't', William J. Broad, *The New York Times*, 16 July 2019, https://www.nytimes.com/2019/07/16/science/5g-cellphones-wireless-cancer.html

'Mast fire probe amid 5G coronavirus claims', BBC, 4 April 2020, https://www.bbc.com/news/uk-england-52164358

'Putin's long war against American science', William J. Broad, *The New York Times*, 13 April 2020, https://www.nytimes.com/2020/04/13/science/putin-russia-disinformation-health-coronavirus.html

'How the 5G coronavirus conspiracy theory tore through the internet', James Templeton, *Wired*, 6 April 2020, https://www.wired.co.uk/article/5g-coronavirus-conspiracy-theory

'The wildest 5G conspiracy theories explained – and debunked', Christian de Looper, 8 April 2020, https://www.digitaltrends.com/news/5g-conspiracy-theories-debunked/

PICTURE CREDITS

All collages and illustrations by Pilar Costabal and some diagrams by Lisa Reidy, except where listed otherwise. All images of Dr Karl Kruszelnicki by Steve Baccon. NOTE: While efforts have been made to trace and acknowledge all copyright holders, in some cases these may have been unsuccessful. If you believe you hold copyright in an image, please contact the publisher.

References are to page numbers.

Coffee – Grinding the Perfect Cup
7   First droplet of an espresso being extracted by radu984/iStock
8   Closeup of coffee beans by Lotus_studio/ Shutterstock.com
8, 9 Foam (crema) on top of hot coffee by Thodsaphol Tamklang/Shutterstock.com
10   Barista making coffee by bunyarit/Shutterstock.com

Dead Fish Swim
14   *Top:* Young humpback whale by Michael Smith ITWP/Shutterstock.com
14   *Centre:* Zebra angelfish (Pterophyllum scalare) in aquarium by juancajuarez/Shutterstock.com
15   *Top:* Lateral line, Creative Commons, https:// en.wikipedia.org/wiki/Lateral_line
15   *Bottom:* Lateral line organ, Creative Commons, https://en.wikipedia.org/wiki/Lateral_line#/media/ File:LateralLine_Organ.jpg
15   *Right border:* School of big-eye trevally jack (Caranx sexfasciatus) by Leonardo Gonzalez/ Shutterstock.com
16   *Top:* Bottlenose dolphin leaping behind a boat by PZ Photos/Shutterstock.com
16   *Bottom:* Underwater picture of female swimmer by LightField Studios/Shutterstock.com

Self-Repairing Lungs
19   Man's hand crushing cigarettes by Aleksandra Voinova/Shutterstock.com
20   3D medical animation of bronchial airways, Creative Commons, https://commons.wikimedia. org/wiki/File:3D_Medical_Animation_Bronchial_ Airways_terminating_ends.jpg
21   Woman breaking cigarette by Dmytro Zinkevych/ Shutterstock.com
22   *Top:* Eukaryote DNA, Creative Commons, https:// commons.wikimedia.org/wiki/File:Eukaryote_ DNA-en.svg
22   *Bottom:* DNA chemical structure, Creative Commons, https://commons.wikimedia.org/wiki/ File:DNA_chemical_structure.svg
23   Embryonic stem cell colony by nobeastsofierce/ Shutterstock.com

Easter & Equinox
25   Easter egg 3D render by lacostique/Shutterstock. com
26   *Top left:* Vernal Equinox, https://publicdomainclip-art. blogspot.com/2017/03/vernal-equinox-ostara.html
26   *Bottom left:* Easter eggs by corradobarattaphotos
27   *Top right:* March 2019: Vernal Equinox, NOAA Environmental Visualization Laboratory, https:// www.nesdis.noaa.gov/content/goes-east- captures-view-vernal-equinox
27   *Centre:* Diagram of Earth's seasons, Creative Commons, https://en.wikipedia.org/wiki/Season#/
28   *Top:* Desk calendar with days and dates in July 2016 by andy0man/Shutterstock.com
28   *Bottom:* 20 April 2019: Cardinals attend the Easter vigil in the Vatican City by Riccardo De Luca - Update/Shutterstock.com

Spiders Can Fly
31   *Top:* Two male Erigone spiders on a grass seed head by Rothamsted Research
31   *Bottom:* Ballooning spiderling, Creative Commons, https://en.wikipedia.org/wiki/ Ballooning_(spider)
32   'An observational study of ballooning in large spiders: Nanoscale multifibres enable large spiders' soaring flight' by M. Cho, P. Neubauer C. Fahrenson, I. Rechenberg (2018), courtesy of Dr Moonsung Cho, https://journals.plos. org/plosbiology/article?id=10.1371/journal. pbio.2004405
35   Courtesy of Dr Erica Morley, 'Electric fields elicit ballooning in spiders', Erica L. Morley and Daniel Robert, Current Biology, 23 July 2018, pp 2324–2330

Past Plagues & Coronavirus
37   Males with face masks by Lek in a BIG WORLD/ Shutterstock.com
38   Little girl wearing mask for protection against COVID-19, MIA Studio/Shutterstock.com
39   Plague in an ancient city, Creative Commons, https://en.wikipedia.org/wiki/Plague_of_Athens#/
40   Transmission electron micrograph (TEM) depicting a number of smallpox virus virions by CDC/ Dr. Fred Murphy/Sylvia Whitfield, Centers for Disease Control and Prevention's Public Health Image Library, https://phil.cdc.gov/Details. aspx?pid=1849
41   *Top:* Artwork of Australian colonial soldiers visiting Botany Bay by Simon Stone/Alamy
41   *Centre:* 1764: Pontiac, an Ottawa Indian, confronts Colonel Henry Bouquet by MPI/Getty Images
41   *Bottom left:* Variola lesions on chest and arms, Creative Commons, https://commons.wikimedia. org/wiki/Category:Smallpox#/
41   *Bottom right:* Smallpox viruses, a colourised transmission electron micrograph by Everett Collection/Shutterstock.com
42   *Top:* COVID-19 disease prevention, aircraft interior cabin deep cleaning by Pradpriew/ Shutterstock.com
42   *Bottom:* Smallpox introduced into Mexico by the Spanish expedition of Panfilo de Narvaez by Everett Collection/Shutterstock.com
43   Spread of Bubonic plague in Europe, Creative Commons, https://en.wikipedia.org/wiki/Black_ Death_migration#/

Red Sky at Night
45   Earth global circulation, Creative Commons, https://en.wikipedia.org/wiki/Hadley_cell#/
46   Sir Isaac Newton examining nature of light with the aid of a prism by Everett Historical/ Shutterstock.com
47   Hands make heart sign at gay pride parade by lazyllama/Shutterstock.com
48   Sunrise at Khong Chiam, Ubon Ratchathani by jackapong/Shutterstock.com
49   Young woman with toddler daughter at a picnic by Evgeny Haritonov/Shutterstock.com

Black Holes – Close & Missing
51   *Top:* Landscape with Milky Way by Denis Belitsky/ Shutterstock.com
51   *Bottom:* Looks like an eye or an iris in the sky by Maxal Tamor/Shutterstock.com (elements of this image furnished by NASA)
52   Cygnus X-1, Creative Commons, https://commons. wikimedia.org/wiki/File:Cygnus_X-1.png
53   *Top:* Computer-generated black hole swallowing galaxy by oorka/Shutterstock.com
53   *Bottom:* Water funnel by Shvaygert Ekaterina/ Shutterstock.com
54   HR 6819 hierarchical triple star system, Creative Commons, https://en.wikipedia.org/wiki/HR_6819#/
56   *Top:* 12 December 2006: Professor Stephen Hawking visiting Tel Aviv university by The World in HDR/Shutterstock.com
56   *Centre:* 13 June 2017: The 2017 Nobel laureate, Kip Thorne, giving a talk at the Technion by The World in HDR/Shutterstock.com
56   *Bottom:* Beautiful night sky, the Milky Way and the trees by Triff/Shutterstock.com (elements of this image furnished by NASA)
57   Orbital diagram of Planet Nine, Creative Commons, https://en.wikipedia.org/wiki/Planet_Nine#/

Coronavirus & Copper
58   Copper mine, Atalaya, Spain by Denis Zhitnik/ Shutterstock.com
59   *Top:* The Magura cave in Bulgaria by Ongala/ Shutterstock.com
59   *Centre:* Native copper, Creative Commons, https://en.wikipedia.org/wiki/Copper#/
59   *Bottom:* World production trend of copper, Creative Commons, https://en.wikipedia.org/wiki/Copper#/
60   *Background:* Variety of ancient Chinese coins by TonyV3112/ Shutterstock.com
60   *Top:* The god Horus offers life to the king, Ramesses II, Creative Commons, https:// en.wikipedia.org/wiki/Ankh#/
61   *Bottom left:* A plant for the recycling of home appliances by Lamzinvnikola/Shutterstock.com
61   *Bottom right:* Industrial plumbing by Octavian Lazar/Shutterstock.com
62   *Top:* The east tower of the Royal Observatory, Edinburgh, Creative Commons, https:// en.wikipedia.org/wiki/Copper#/
62   *Bottom:* Nurse makes hospital bed by Sean Locke Photography/Shutterstock.com

The Amazing Disappearing Anus
65   Yellow polyps of coral by Konstantin Novikov/ Shutterstock.com
66   A bootlace worm (Lineus longissimus), Creative Commons, https://en.wikipedia.org/wiki/ Nemertea#/
67   Pelagic ctenophores, Creative Commons, https:// en.wikipedia.org/wiki/Ctenophora#/
68   *Top:* Underwater creatures (Mnemiopsis leidyi) close to the surface of the Caribbean sea by Damsea/Shutterstock.com
68   *Bottom:* Courtesy of Dr Sidney Tamm, 'Defecation by the ctenophore Mnemiopsis leidyi with an ultradian rhythm through a single transient anal pore' by Sidney L. Tamm, Invertebrate Biology, 22 February 2019; 138, pp 3–16
70,71 Courtesy of Dr Camilo Mattoni, 'Scorpion sheds "tail" to escape: consequences and implications of autotomy in scorpions (Buthidae; Ananteris)', Camilo I. Mattoni et al., PLOS One, 28 January 2015, DOI:10.1371/journal.pone.0116639

Murder Hornets – Lethal but Tasty?
73   *Top:* Isolated sting of a hornet by CarlosR/ Shutterstock.com
73   *Centre:* Male Japanese giant hornet (Vespa mandarinia japonica), Creative Commons, https:// en.wikipedia.org/wiki/Asian_giant_hornet#/
73   *Bottom:* Asian giant hornets in nest by NitayaPhet/Shutterstock.com
74   *Top:* Head of Japanese giant hornet, Creative Commons, https://en.wikipedia.org/wiki/Asian_ giant_hornet#/
74   *Bottom:* Honeybee thermal defence, Creative Commons, https://commons.wikimedia.org/wiki/ File:Honeybee_thermal_defence01.jpg
75   Gypsy moth on clematis seed by Alexander Sviridov/Shutterstock.com
76   Green hornet sushi roll by IcemanJ/iStock
77   *Centre:* European hornet (Vespa crabro) by Dmitry Dolhikh/Shutterstock.com
77   *Bottom right:* Mosquito by phichak/Shutterstock.com
77   *Bottom left:* Hand with a swollen large middle finger by Inna Kozhina/Shutterstock.com

Women's Work – Never Done, Never Paid
79 14 June 2019: Women's strike in the streets of Zurich by elizabethdalessandro/Shutterstock.com
80 *Top:* Young teacher with toddlers by Krakenimages.com/Shutterstock.com
80 *Bottom:* Women carrying water pots on their heads by SkycopterFilms Archives/Shutterstock.com
81 *Top:* 4 September 1991: Warren Buffett testifies before US House subcommittee by Rob Crandall/Shutterstock.com
81 *Centre:* Nurse helping mature patient by Dmytro Zinkevych/Shutterstock.com

Black Holes Have No Size
85 Interplanetary station Luna 1, Creative Commons, https://en.wikipedia.org/wiki/Luna_1#/
86 Visual wavelength image of Messier 87, Creative Commons, https://en.wikipedia.org/wiki/Messier_87#/
87 The M87 jet seen by the Very Large Array in radio frequency, Creative Commons, https://en.wikipedia.org/wiki/Astrophysical_jet#/
88 Black hole, Creative Commons, https://commons.wikimedia.org/wiki/File:Black_hole_-_Messier_87_crop_max_res.jpg

Pterosaurs, Not Just Pterodactyls
91 Rhamphorhynchus muensteri, Creative Commons, https://commons.wikimedia.org/wiki/File:Cast_of_Rhamphorhynchus_muensteri_02_-_Pterosaurs_Flight_in_the_Age_of_Dinosaurs.jpg
92 *Top:* Artist's impression of a Quetzalcoatlus in flight, Creative Commons, https://en.wikipedia.org/wiki/Quetzalcoatlus#/
92 *Centre:* 3 May 2015: Jet fighter F-16 block 52 plus demo team Zeus in an air show by thelefty/Shutterstock.com
92 *Bottom:* Comparison of Quetzalcoatlus northropi and Cessna 172 light aircraft, Creative Commons, https://en.wikipedia.org/wiki/Quetzalcoatlus#/
93 *Centre:* Skeletal reconstruction of a quadrupedally launching Pteranodon, Creative Commons, https://commons.wikimedia.org/wiki/File:Quad_launch.jpg
93 *Bottom right:* Artist's impression of a group of Quetzalcoatlus feeding, Creative Commons, https://en.wikipedia.org/wiki/Quetzalcoatlus#/

Marathon Runners' Gut Bacteria
95 *Top:* Woman running by lzf/Shutterstock.com
95 *Bottom:* Stem cells diagram, Creative Commons, https://commons.wikimedia.org/wiki/File:Stem_cells_diagram.png
96 Marathon runners by catwalker/Shutterstock.com
99 Layers of the alimentary canal, Creative Commons, https://en.wikipedia.org/wiki/Gastrointestinal_tract#/

Combustion Engines – The Burners Go Bust?
101 A 4WD vehicle attempts to climb the 'Big Red' sand dune in outback Australia by Ian Hitchcock/Shutterstock.com
102 *Top:* CG model of a working V8 engine with explosions and sparks by yucelyilmaz/Shutterstock.com
102 *Bottom:* Bertha Benz portrait, Creative Commons, https://commons.wikimedia.org/wiki/File:Berthabenzportrait.jpg
103 Vintage black steam train by Arcansel/Shutterstock.com
104 *Top:* 1886 Benz Patent-Motorwagen, the first car by Karl Benz by Rudiecast/Shutterstock.com
104 *Bottom:* Charging modern electric car on the street by guteksk7/Shutterstock.com
105 New Tesla car in factory by Nadezda Murmakova/Shutterstock.com

Time is Running Out for Sand
107 Dubai skyline in desert at sunset by WaitForLight/Shutterstock.com
108 *Top right:* Portrait of Amedeo Avogadro, Creative Commons, https://commons.wikimedia.org/wiki/Amedeo_Avogadro
108 *Bottom left:* Open-pit mining and processing plant for crushed stone, sand and gravel by Dmitry Rukhlenko/Shutterstock.com
109 Shells and pebbles on beach sand by Feel good studio/Shutterstock.com
110 *Top:* Skyline view of Dubai Mall, Dubai Fountain and the Burj Khalifa by Sophie James/Shutterstock.com
110 *Bottom:* High-resolution satellite image of Dubai coast from above by TommoT/Shutterstock.com
111 *Bottom:* Sand use in Singapore, UNEP/GRID-Geneva, https://na.unep.net/api/geas/articles/getArticleHtmlWithArticleIDScript.php?article_id=110
111 *Top:* 31 August 2012: Anouk Verge-Depre competes at the FIVB Beach Volleyball Swatch Junior World Championships by Jamie Roach/Shutterstock.com
113 Redrawn by Pilar Costabal from 'Direct and indirect impacts of aggregate dredging, Marine Aggregate Levy Sustainability Fund (MALSF)', H.M. Tillin, A.J. Houghton, J.E. Saunders, R. Drabble and S.C. Hull, Science Monograph Series 1, 2011, pp 1–46

Vegan Diet
115 Healthy vegetarian food by nadianb/Shutterstock.com
117 White lambs by aaltair/Shutterstock.com
118 Homemade roasted brussel sprouts by nelea33/Shutterstock.com
119 Veggie beet and carrot burgers with avocado by Kolpakova Svetlana/iStock

Fisher Space Pen – Out of This World!
121 Astro-space pen, Creative Commons, https://commons.wikimedia.org/wiki/File:Astro-space-pen.jpg
122 Newspaper clipping courtesy of NASA
123 US patent drawing (public domain)

Fish Exercising
127 *Enzo the Wonderfish*, Cathy Wilcox, Angus & Robertson, 1993
128 Water vortex by block23/Shutterstock.com
128 Courtesy of Prof Takuji Usui & Prof Shinichi Nakagawa, 'Using a French Press Coffee Maker to Exercise Fish', Journal of Improbable Research

Surfing Space-Time – Einstein's Right Again!
132 19 October 2019: Miriam Abdulkarimova performs ladies free skating program in Ice Star championship by Dmitry Morgan/Shutterstock.com
133 *Top:* Fred Astaire, Ginger Rogers by Courtesy Everett Collection/Alamy
133 *Centre:* Artist's impression of the WDJ0914+1914 system, Creative Commons, https://commons.wikimedia.org/wiki/File:Artist%E2%80%99s_impression_of_the_WDJ0914%2B1914_system.tif
133 *Bottom:* Schematic view of a pulsar, Creative Commons, https://en.wikipedia.org/wiki/Pulsar#/
135 The perihelion precession of Mercury, Creative Commons, https://en.wikipedia.org/wiki/Tests_of_general_relativity#/
137 Portrait of Albert Einstein by Library of Congress/alamy

Carbs & Ultra-Processed Foods
141 *Centre:* A 3D schematic representation of a sucrose molecule, Creative Commons, https://en.wikipedia.org/wiki/Disaccharide#/
141 *Bottom:* View of the atomic structure of a single branched strand of glucose, Creative Commons, https://en.wikipedia.org/wiki/Polysaccharide#/
142 *Top:* Schematic 2D cross-sectional view of glycogen, Creative Commons, https://en.wikipedia.org/wiki/Polysaccharide#/
142 *Bottom:* Sugar by margouillat photo/Shutterstock.com
143 *Top:* Production line of chocolate cookies by mady70/Shutterstock.com
143 *Centre:* Assorted organic craft sodas by Brent Hofacker/Shutterstock.com
144 Crispy potato chips in a wicker bowl by Jiri Hera/Shutterstock.com
145 *Top:* Worker using machine in bakery factory by Travelerpix/Shutterstock.com
145 *Centre:* Overweight child on scale by Eviart/Shutterstock.com

Spiders Can Design, Build – And Count!
147 The spinneret of an Australian garden orb weaver spider, Creative Commons, https://commons.wikimedia.org/wiki/File:Australian_garden_orb_weaver_spider_spinneret.jpg
148 A cobweb with dew drops strung between branches by caitrionad /iStock
148 Close-up macro of spider on web by nate samui/Shutterstock.com
149 Spinnerets of female Eastern Parson Spider, Creative Commons, https://commons.wikimedia.org/wiki/File:Spinnerets.jpg
151 A Cross spider weaving a new orb web by marthadavies/iStock

Winded – A Breathtaking Experience
153 *Top:* Man suffering from stomach pain by WSW1985/Shutterstock.com
153 *Bottom:* Structure of diaphragm using 3D medical animation, Creative Commons, https://en.wikipedia.org/wiki/Thoracic_diaphragm#/
154 X-Ray image of human chest, tuberculosis screening by memorisz/Shutterstock.com
155 Fit man in sportswear by mavo/Shutterstock.com

Migrating Planets – A Different Tack
158 Bronze statue of Lord Kelvin in Kelvinside, Glasgow, Creative Commons, https://commons.wikimedia.org/wiki/File:Bronze_statue_of_Lord_Kelvin_in_Kelvinside,_Glasgow_-_geograph.org.uk_-_1592023.jpg
160 Solar system with stars by Ezume Images/Shutterstock.com
163 Solar system with an asteroid belt and a comet, NataliaMalc/Shutterstock

5G Mania
165 Nomad mother on cell phone in Sahara desert by cdrin /Shutterstock.com
167 The electromagnetic spectrum, Creative Commons, https://en.wikipedia.org/wiki/Electromagnetic_spectrum#/
168 Telecommunication tower with antenna on city background by Suwin/Shutterstock.com
169 Procedure for a mobile phone cooking an egg supplied by Dr Karl Kruszelnicki, https://www.abc.net.au/science/articles/2007/08/23/2012756.htm
173 16 April 2018: Bill Gates at the Élysées Palace by Frederic Legrand, COMEO/Shutterstock.com
174 *Top:* Woman smears face sunscreen at the beach for protection by Tymonko Galyna/Shutterstock.com
174 *Centre:* Tan on shoulder of a child by Irishasel/Shutterstock.com
174 *Bottom:* 5G antenna mast by JazzLove/Shutterstock.com
177 Portrait of Vladimir Putin by Erkan Atay/Shutterstock.com
178 Ultrasound therapy by Leonid Salkhin/Shutterstock.com
179 Space satellite monitoring weather by Andrey Armyagov/Shutterstock.com (elements of this image furnished by NASA)

THANK YOU!

The universe is an amazing place. We are so lucky to be alive and enjoying all it has to offer! Awareness of the big picture only adds to the pleasure – and to the responsibility we have to protect it all.

Scientists continually explore both the problems and the wonders of the universe. Their discoveries wind up in popular science journals, which I read to glean stories that pique my interest, and then I retell these important discoveries in plain English for a broader audience! I am so grateful to all these scientists and science journalists.

Thanks also to all the people in my life who help get my work 'out there'. My wife, Mary Dobbie, helps edit and polish my initial drafts into stories. The science producers/journalists at the ABC (Bernie Hobbs and Carl Smith) turn my work into radio stories. Isabelle Benton (my producer) comes up with lovely punchlines and emanates calm, can-do vibes and good will.

I know that I am a generalist, and not an expert in any single field. So, aiming for accuracy, I run my stories past experts in relevant fields. Specifically, my thanks go to Professor Clare Collins on dietary/foody stories, and Professor Geraint Lewis for astronomy/cosmology/quantum physics stories. I also thank the following scientists for helping me with stories relating to their particular research – Dr Sidney Tamm, Professor Camilo Mattoni and Dr Phil Barnes ('The Amazing Disappearing Anus'); Prof. Shinichi Nagakawa ('Fish Exercising'); and Dr Erica Morley and Dr Moonsung Cho ('Spiders Can Fly').

Chris Steadman reads every word on every page and does the best combination of fact-/grammar-/reference-checking I have ever seen.

At HarperCollins, I was fortunate to work very closely with Lu Sierra, who went above and beyond in the complex task of turning simple text and images into this beautifully integrated final version. Designer Lisa Reidy and illustrator Pilar Costabal collaborated to create the lovely look and feel of this book. The design aesthetics of the book are everything I could have hoped for – colourful, quirky and impressive. Scott Forbes, Nikki Lusk and Annabel Adair respectively in-house coordinated, copyedited and proofread magnificently.

For the hologram/augmented reality features, almost-Doctor Petr Lebedev videoed the pop-ups of me – he's a natural multi-tasker. Shayne Reynolds from AuggD very nicely integrated the entire hologram/augmented reality experience.

I thank my publisher, Jude McGee, who had the vision; my agent Jeanne Ryckmans; Brendan Fredricks (publicity); Alice Wood, Tom Saras and Camellia Cratt (marketing); and Steve Baccon (photographer).

I just love how this book turned out, and I hope that you will also.

ABOUT DR KARL

Dr Karl Kruszelnicki was lucky enough to live in a time when the Australian government saw education as a worthwhile investment in the future — and so he has been the beneficiary of 28 years of free education. That includes 16 years at various universities studying degrees in physics, mathematics, biomedical engineering, medicine and surgery, as well as non-degree years in astrophysics, electrical engineering, computer science and philosophy to round him off. So, of course, he wants to balance the books and give back by sharing his knowledge and love of learning. For four decades, Dr Karl has spread the good word on Science via radio, TV, internet and print, and this is his 46th book.

Besides working as a 4WD test driver for 20 years, labourer, university academic, physicist, roadie for Bo Diddley, taxi driver, TV weatherman, medical doctor, etc, Dr Karl also made some of the first music videos for Australian TV, and helped set up Australia's first cable TV network at the Nimbin Aquarius Festival in 1973.

Since 1995, Dr Karl has been the Julius Sumner Miller Fellow at the University of Sydney.

In 2019, he was awarded the UNESCO Kalinga Prize for the Popularisation of Science, previous recipients of which include Margaret Mead, David Attenborough, Bertrand Russell and David Suzuki.

Every week, Dr Karl gives free Science Q&A sessions with schoolchildren around the world. To book a Skype/Zoom/Teams session for your school, go to drkarl.com.

ABC BOOKS

The ABC 'Wave' device is a trademark of the
Australian Broadcasting Corporation and is used
under licence by HarperCollins*Publishers* Australia.

First published in Australia in 2020
by HarperCollins*Publishers* Australia Pty Limited
ABN 36 009 913 517
harpercollins.com.au

HarperCollins*Publishers*
Level 13, 201 Elizabeth Street, Sydney, NSW 2000, Australia
Unit D1, 63 Apollo Drive, Rosedale, Auckland 0632, New Zealand
A 53, Sector 57, Noida, UP, India
1 London Bridge Street, London, SE1 9GF, United Kingdom
Bay Adelaide Centre, East Tower, 22 Adelaide Street West, 41st Floor, Toronto, Ontario, M5H 4E3, Canada
195 Broadway, New York, NY 10007, USA

A catalogue record for this book is available from the National Library of Australia

ISBN 978 0 7333 4033 8 (paperback)
ISBN 978 1 4607 1167 5 (ebook)

Cover design by Lisa Reidy
Cover images: collage by Pilar Costabal; Dr Karl by Steve Baccon; images by istockphoto.com
Internal design by Lisa Reidy
Colour reproduction by Splitting Image
Printed and bound in China by RR Donnelley